Freehand Sketching

SECOND EDITION

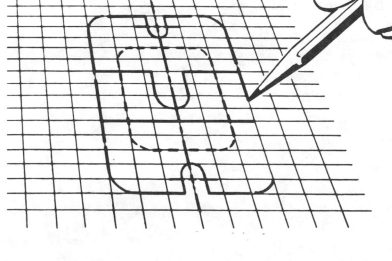

AMERICAN TECHNICAL PUBLISHERS, INC.
HOMEWOOD, ILLINOIS 60430

LIBRARY OF CONGRESS CATALOG CARD NUMBER: 72-89959

ISBN: 0-8269-1022-X

Preface to the Second Edition

The important skill of freehand sketching can be developed readily by following the course of study in this text.

In this Second Edition of this very popular text, the original eight chapters are revised to update them and make them even more effective for quicker learning.

Two entirely new chapters are added to extend the coverage and to add to the planned practice for developing skill—48 pages in all.

The development of freehand sketching skill by following this course can be most rewarding, for one can often convey a detailed idea better with a sketch rather than words.

The Authors

Contents

UNIT 1

BASIC PRINCIPLES OF FREEHAND SKETCHING

At one time or another, in your work or recreation, you will be confronted with the task of making some kind of a sketch. Perhaps the sketch will involve an illustration of the project you plan to make in the shop, or it may portray some repair work you wish to do around the home. It might even be used to convey instructions to another person or to illustrate some written report.

Making freehand sketches is not difficult if you follow a few simple rules. Naturally, your progress may be slow to start with, but after a little practice you will be able to turn out reasonably good sketches without too much effort. In this unit you will find the basic principles you will need to know in making freehand sketches.

Paper for Sketching. Experienced draftsmen use plain, unruled paper for sketching. As a beginner, it is advisable to practice your sketching on squared or cross-section paper. With this kind of paper you will find it easier

to draw straight lines and secure good proportions. Squared paper also simplifies laying out exact sizes without using a scale or rule. Thus, if the paper has squares of a consistent size, each square can represent $1/4''$, $1/2''$ or any other desired size (Fig. 1).

Proportions. Maintaining correct proportions is a very important feature in freehand sketching. Ordinarily, regardless of how excellent the sketching techniques may be, the final sketch itself will not be a very good one if it possesses relatively poor proportions.

Proper proportioning is achieved by estimating actual dimensions. Sketches are not made to scale and one has to learn to recognize proportions by comparison. Thus you will start by noting the relationship of the width, length and depth. If, for example, the height is twice the width, then this same proportion can be maintained for other details.

The problem of holding proportions is not quite as difficult if grid paper is used. By counting the number of squares, the relationship of sizes can be easily followed. Sometimes the process is simplified if the general

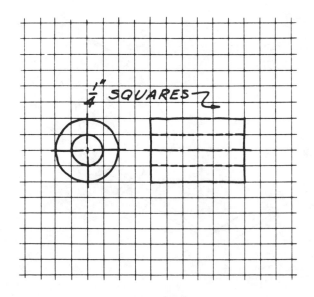

Fig. 1. Note how squared paper simplifies making a sketch.

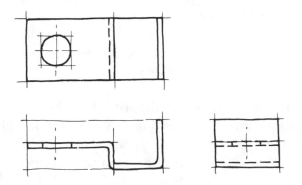

Fig. 2. Sketching rectangles or squares over the area enclosing the views will often help achieve correct proportions.

1

area enclosing a view is divided into squares or rectangles having proportions which are the ones to be maintained for the view. See Fig. 2.

Position of the Pencil. Use either a medium (F) or soft (HB) pencil for sketching. The symbol showing the hardness of the lead is

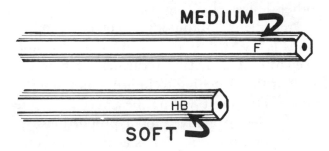

Fig. 3. Use one of these pencils for sketching.

stamped on the pencil (Fig. 3). Sharpen the point in an ordinary pencil sharpener.

Hold the pencil loosely about 1½″ to 2″ away from the point, as in Fig. 4. As the pen-

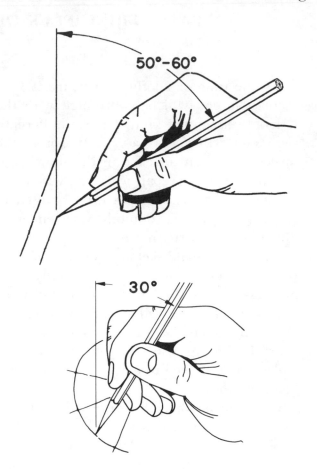

Fig. 5. Slant the pencil at these angles for lines and circles.

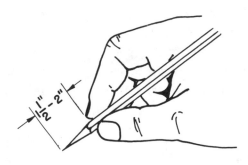

Fig. 4. Hold the pencil in this position for most drawing purposes.

cil is used, rotate it slightly. This keeps the point sharp longer and makes clear lines. The general practice is to slant the pencil at an angle of 50 to 60 degrees from the vertical for drawing straight lines and about 30 degrees for circles (Fig. 5). Some draftsmen hold the pencil in a flat position for straight lines. In this position, the hand is glided on the back of the fingernails (Fig. 6).

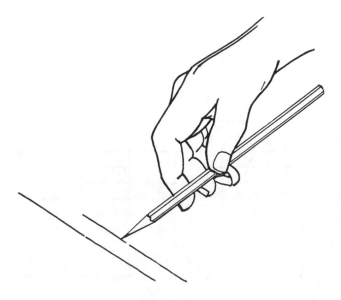

Fig. 6. Some people like to hold the pencil in a flat position for making straight lines.

Always pull the pencil in sketching straight lines and curves. If the pencil is pushed, it may catch the surface of the paper and puncture or tear it.

Sketching Horizontal Lines. To draw horizontal lines, first mark off two points to indicate the position of the lines (Fig. 7). Then sketch the line between the two points, moving the pencil from left to right. For short lines, use a finger and wrist movement. As the line becomes longer, it is better to use a free arm movement, since the fingers and wrist tend to bend the line.

Fig. 7. Mark off the position of the line by two points and sketch line from left to right.

Draw short lines in a single stroke. Long lines are drawn more accurately if they are made in a series of short strokes (Fig. 8). By using short strokes of 1½″ to 2″ long, it is easier to maintain the proper direction. A space of approximately ¹⁄₃₂″ is left between strokes.

Fig. 8. Long lines are sketched more accurately by a series of short strokes.

Sketching Vertical and Slanted Lines. Sketch vertical lines by starting at the top of the paper, drawing the pencil downward. Slanted lines can be sketched better if the pencil is moved from left to right (Fig. 9). Sometimes it is advisable to turn your paper so the vertical or slanted lines assume a horizontal position (Fig. 10).

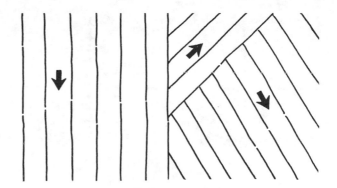

Fig. 9. Sketch vertical and slanted lines in this manner.

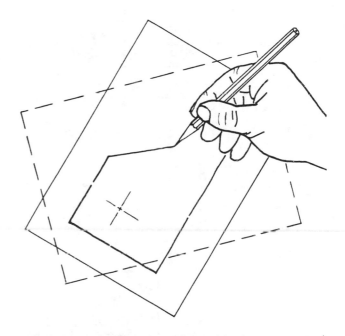

Fig. 10. Turning the paper will simplify drawing vertical slanted lines.

Sketching Figures. Geometric constructions, particularly squares, rectangles, angles, circles, and irregular curves, must be drawn frequently. The following paragraphs outline simple methods of drawing these figures rapidly and accurately.

Squares. To sketch a square, draw a vertical and horizontal line as shown in *A* (Fig. 11). Space off equal distances on these lines as in *B*. Sketch light horizontal and vertical lines through the outer points to form the square. Then darken the lines as in *C*. Another way to get the outer points for the sides of the

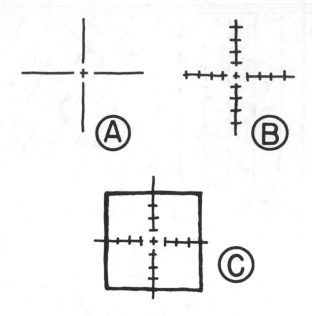

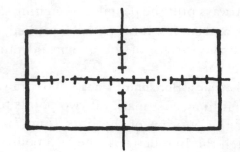

Fig. 13. How to sketch a rectangle.

Angles. To sketch an angle, draw two lines to form a right angle. Divide the right angle into equal spaces. Project a line through the point that represents the angle desired (Fig. 14).

Circles. To sketch a circle, draw a horizontal and vertical line through a point marking

Fig. 11. How to sketch a square.

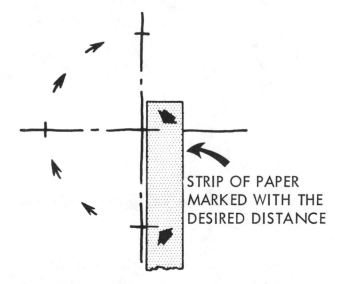

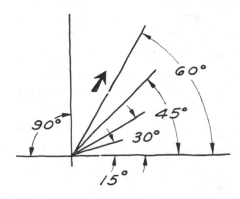

Fig. 14. How to sketch an angle.

Fig. 12. The points for a square are laid out in this manner.

square is to mark them with a piece of scrap paper as shown in Fig. 12.

Rectangles. Sketch two lines for the center of the rectangle. Space off these lines to provide the required width and length of the rectangle. Then draw the necessary lines through the outer points to form the rectangle (Fig. 13).

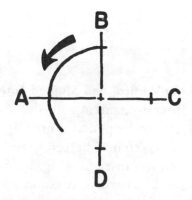

Fig. 15. How to sketch small circles.

the center of the desired circle. Find the outer point of the circle, either by dividing each line into small spaces or by using a scrap paper as described in sketching a square. Complete the circle by sketching short arcs from *B* to *A*, *A* to *D*, *B* to *C*, and *C* to *D* (Fig. 15). For a large circle, it will be helpful if a series of radii are drawn as illustrated in Fig. 16.

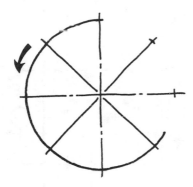

Fig. 16. How to sketch large circles.

Irregular Curves. For an irregular curve, lay out a number of points to represent the required curvature. Complete the curve by sketching a series of arcs through these points (Fig. 17).

Fig. 17. How to sketch an irregular curve.

Sketching a Flat Layout. A flat layout is really a one-view drawing. It is an actual outline of the object to be made (Fig. 18). In shop work, this type of outline must frequently be made. For example, you may want to make a drill gauge, tray, bird house, bracelet, fixture,

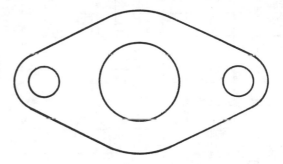

Fig. 18. An example of a flat layout.

or a jig. In each case, it will be necessary to lay out the shape of every piece you intend to make. Before starting the work, you will want to sketch the parts to get an idea of their shape. The common practice is to sketch such parts on paper and then transfer the design on the material to be used.

To make a flat layout, proceed as follows (Fig. 19):

1. Draw base or center lines and locate arcs and circles.
2. Block in the outline of the object with light lines.
3. Draw circles and arcs first, then draw straight lines and darken all outlines.

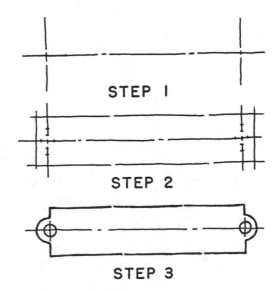

Fig. 19. Steps in sketching a layout.

PROBLEMS 1 and 2. *Make full-size sketch.*

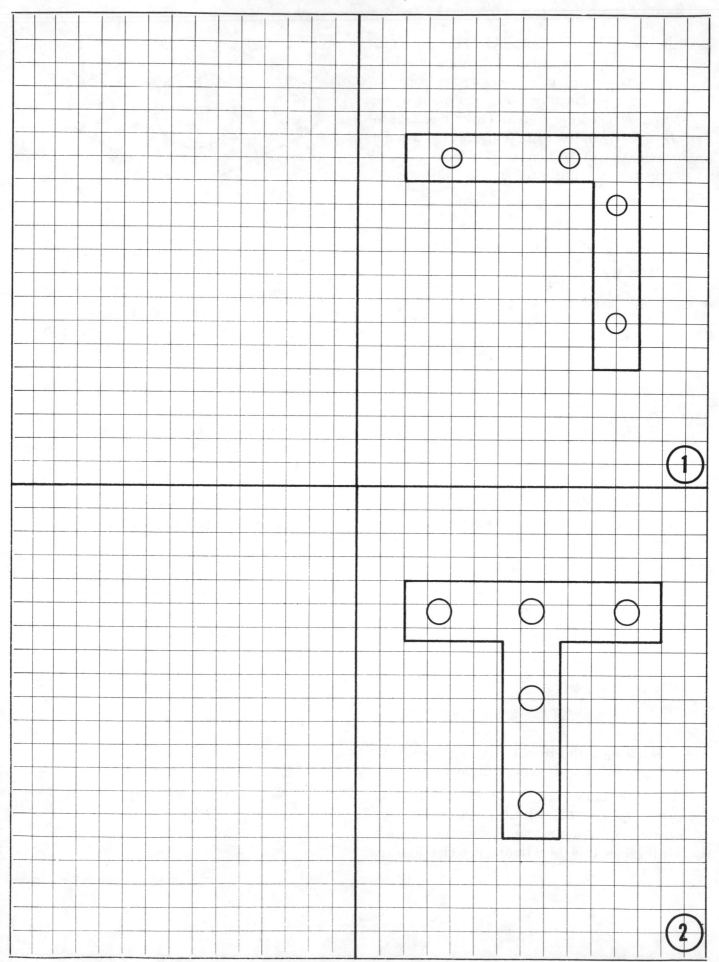

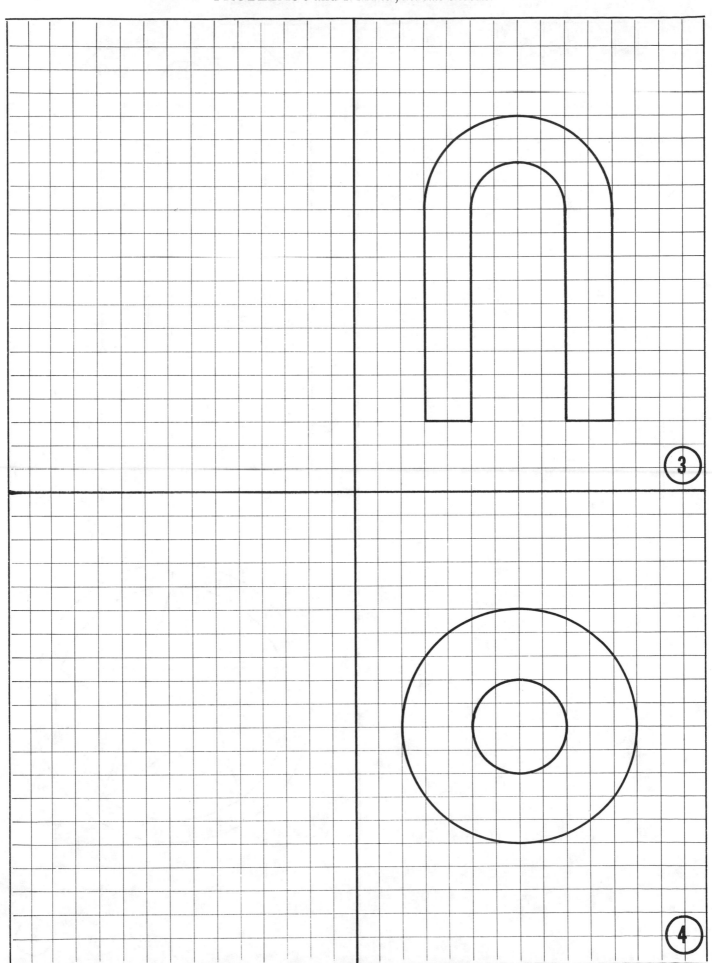

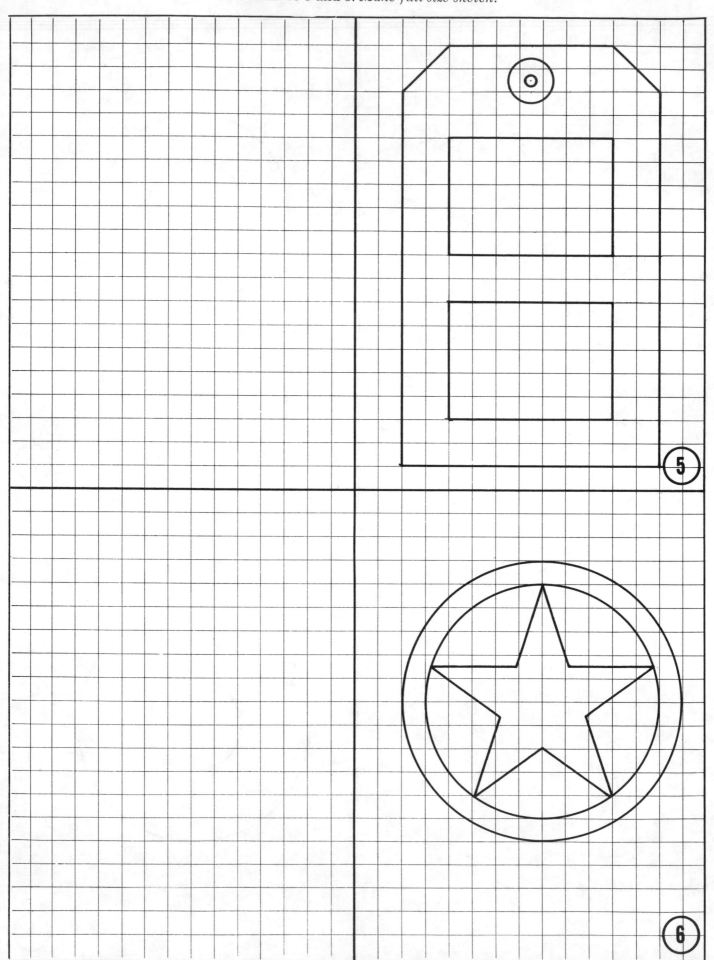

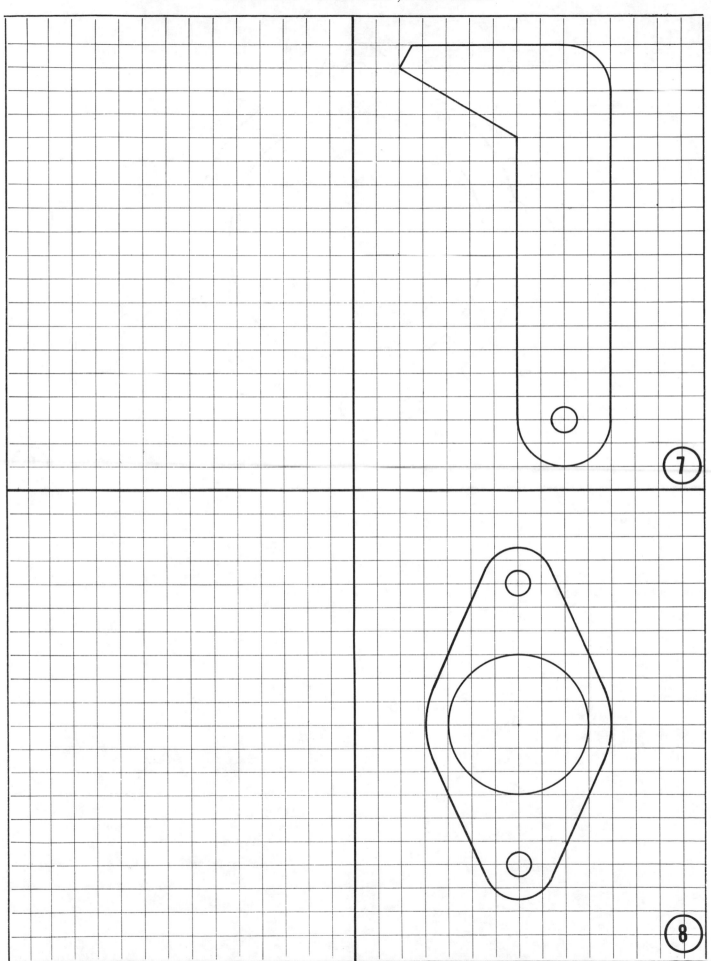

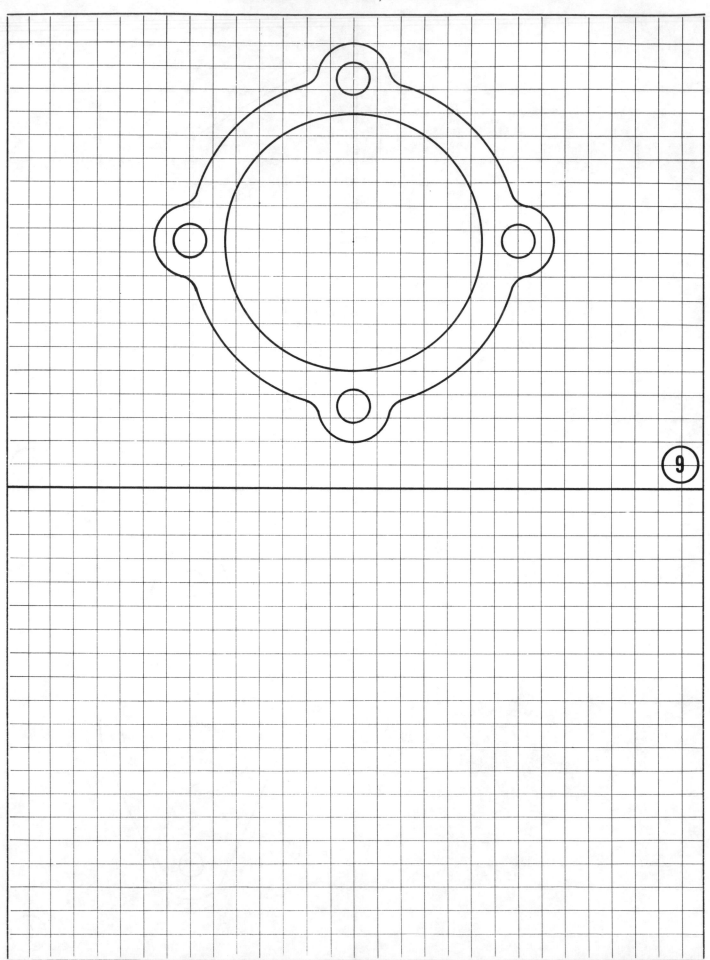

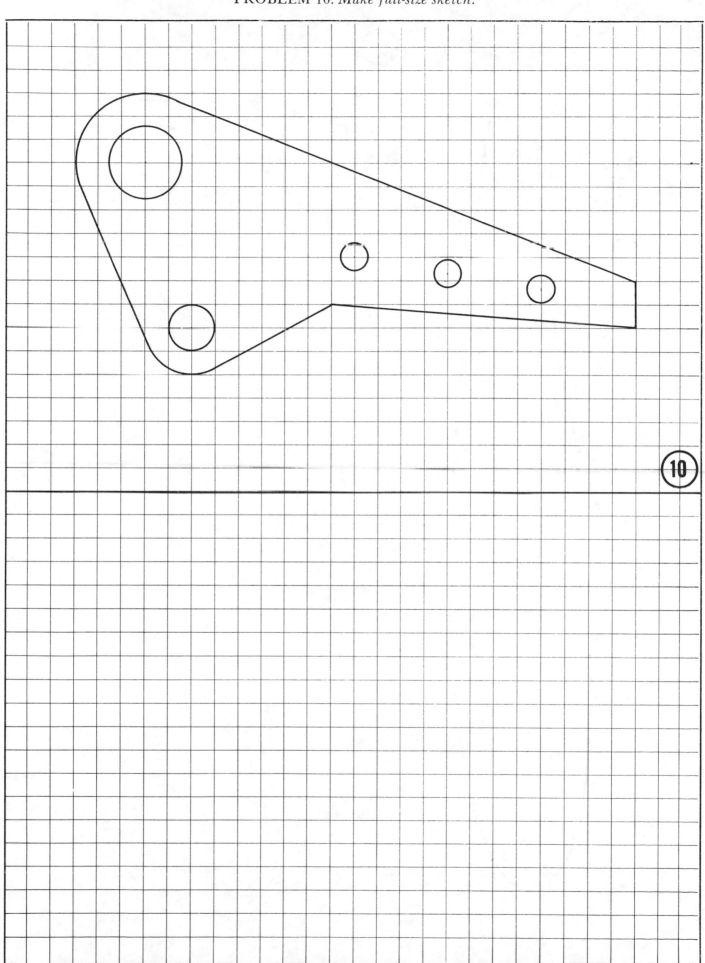

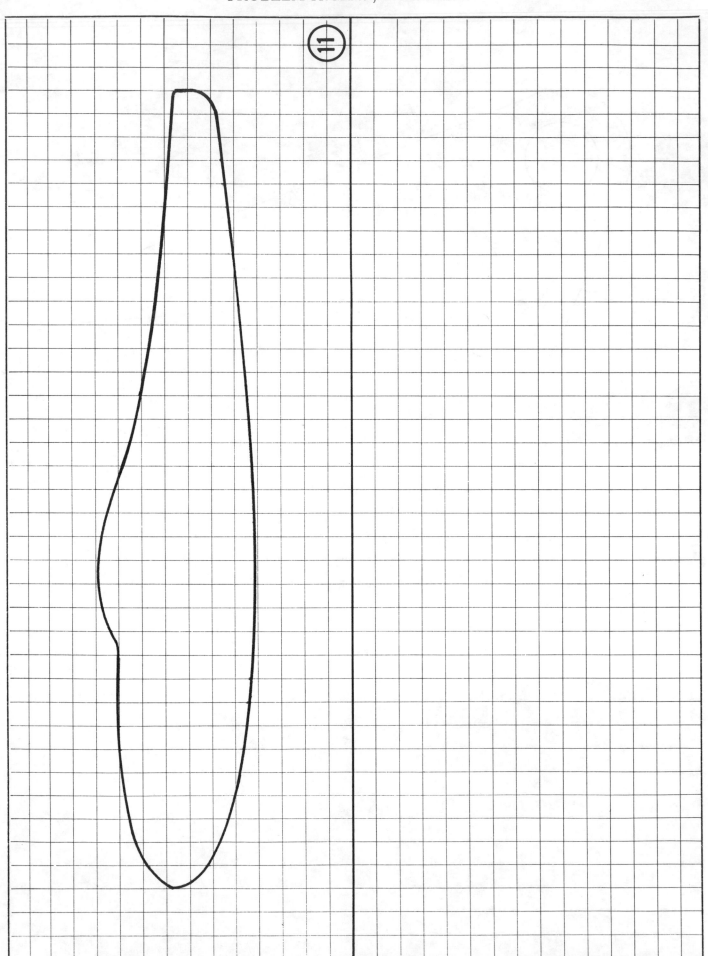

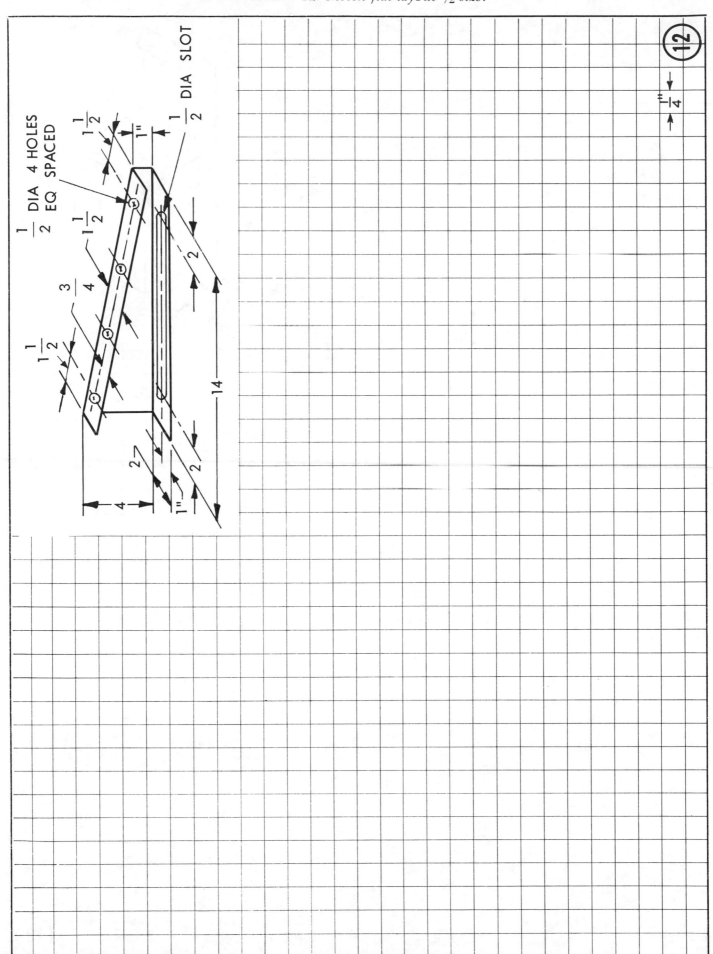

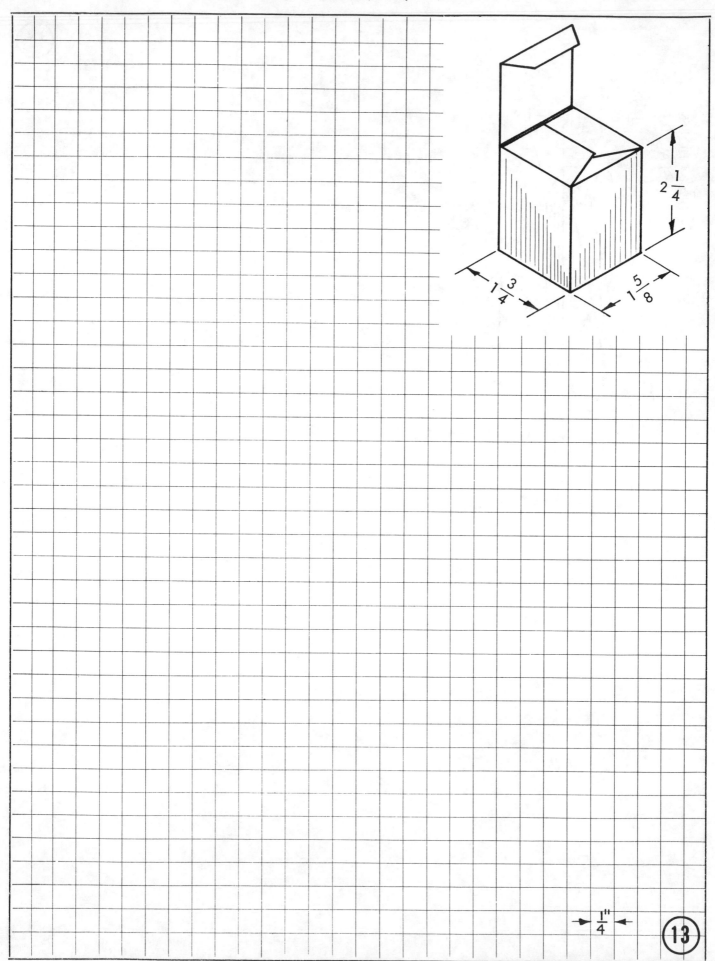

$2\frac{1}{4}$

$1\frac{3}{4}$

$1\frac{5}{8}$

$\frac{1}{4}''$

13

PROBLEM 14. *Sketch half of pattern full size.*

PROBLEM 15. *Make sketch of appropriate size to fit on a Practice Sheet at the end of the book.*

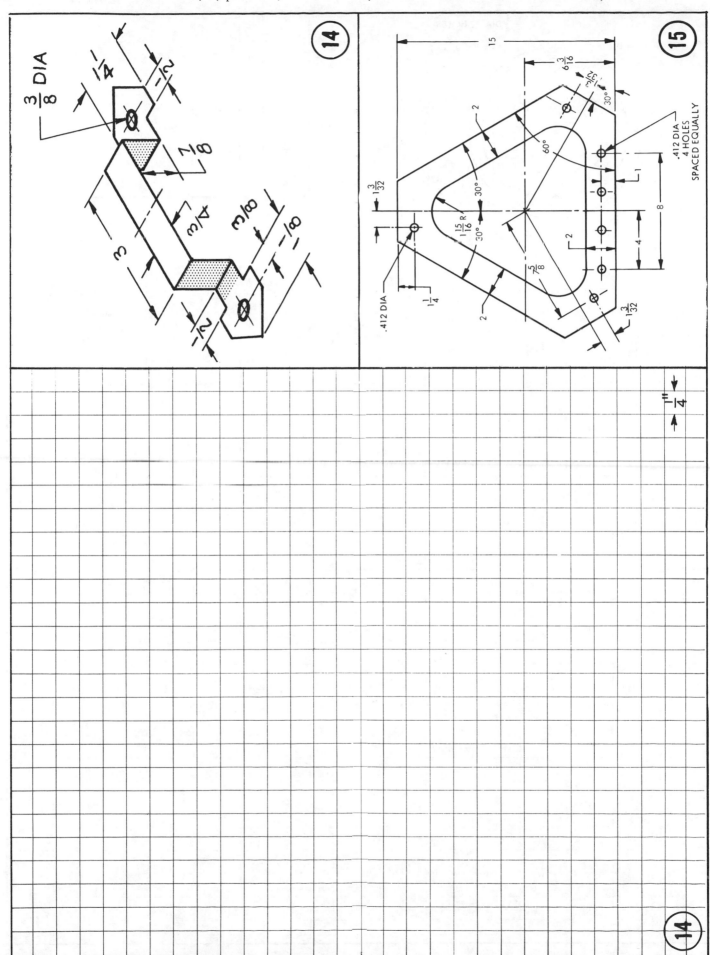

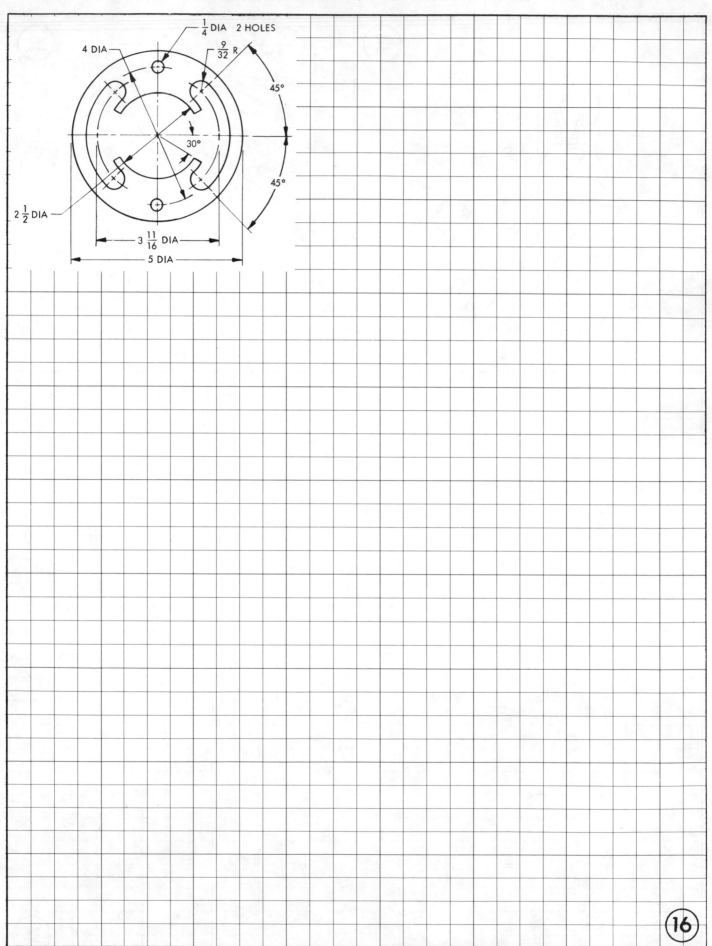

16

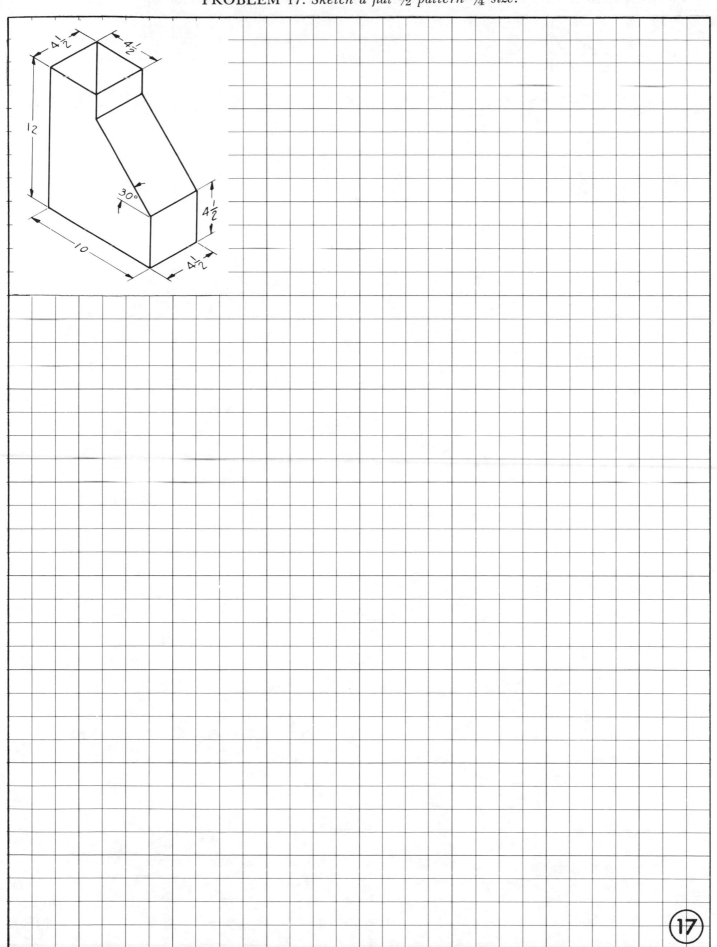

PROBLEM 18. *Sketch a flat layout of appropriate size to fit in sheet. Maintain proper proportions.*

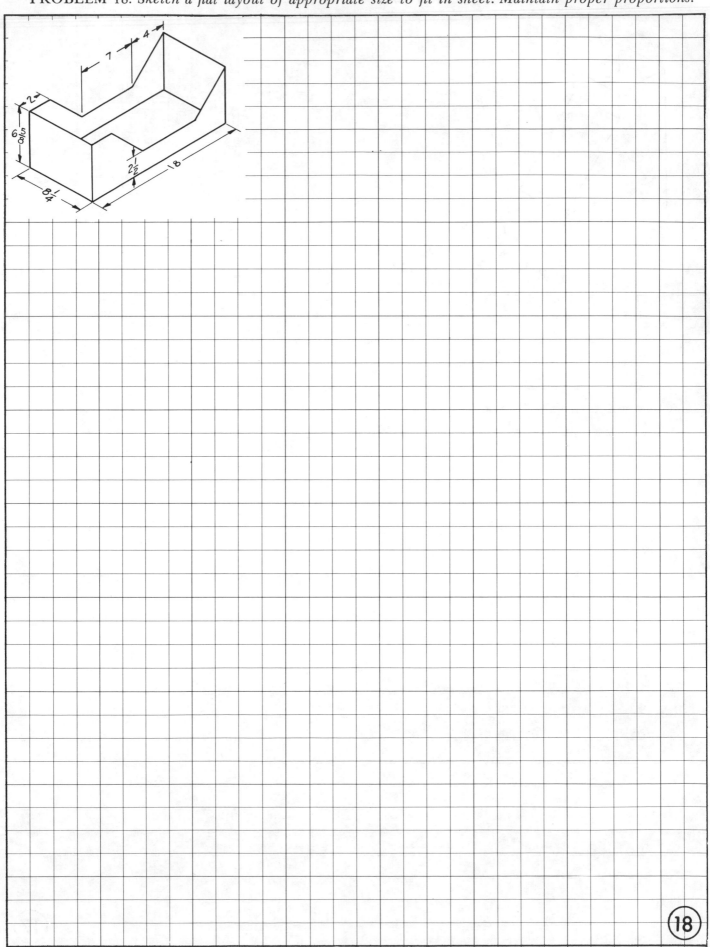

18

UNIT 2

LETTERING A SKETCH

When making a sketch of an object, you may have to add dimensions and other information in the form of letters and notes. If these numbers and letters are to be formed correctly, they must be made in a certain way. By following a few simple rules, and with a little practice, you can quickly master the art of good lettering.

Forming Letters and Numbers. Today the common practice in industry is to use all upper case (capital) letters. They can be either slanted or vertical (Figs. 1 and 2). There is no specific rule as to whether the lettering should be of the slant or vertical type. Insofar as you are concerned, try out both styles. Concentrate, then, on the one that appeals to you more and that you can make more easily. The important thing is to make the letters uniform so they will have a pleasing appearance.

To make attractive letters and numbers, use the strokes as shown in Figs. 1 and 2. You will find that by following these strokes, lettering is really a simple operation.

If you study Fig. 3, you will see that the

Fig. 1. These are vertical letters.

Fig. 2. Slanted letters are formed in this manner.

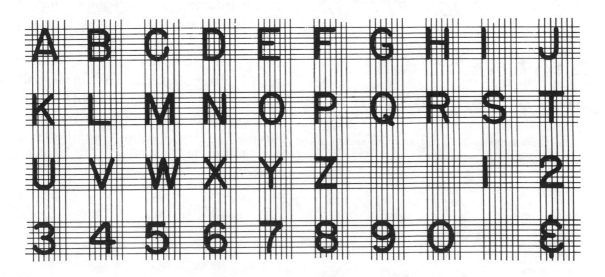

Fig. 3. Correct proportions of letters and numbers.

19

width of letters varies. Thus, some letters are almost as wide as they are high, while others are slightly narrower. To give you a better idea of the relationship between width and height, the letters and numbers in Fig. 3 are reproduced in blocks six squares high and six squares wide. Notice that letters such as B, C, D, E, F, G, etc., are four squares wide. Others, such as A, O, Q, R, and V, are four and one-half squares wide, whereas M is five squares wide. The widest letter of the alphabet is W, which is six squares wide.

Notice, also, the correct proportions for numbers. You will see that most numbers, such as 2, 3, 7, 8, are about two squares narrower than they are high, while 4, 6, 9, and 0 are more than four squares wide. For most sketches, letters and numbers should be about ⅛″ high.

Spacing Letters and Words. Making attractive letters is not enough for proper lettering. Equally important is the spacing between letters. To get a pleasing appearance, the areas between each letter must appear to be equal.

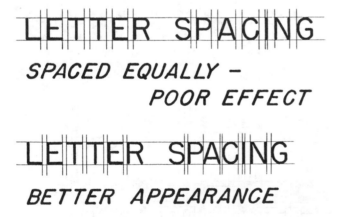

Fig. 4. Notice how these letters are spaced to secure a pleasing appearance.

Due to the shapes of various letters, the appearance of equal areas cannot always be obtained by simply setting the letters a uniform distance apart. Keep in mind that less distance is required between some letters (Fig. 4).

The distance between words should be equal to the height of the letters (Fig. 5). The

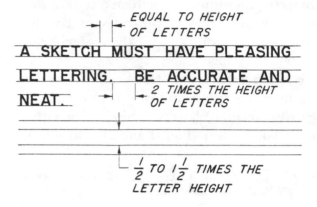

Fig. 5. This is how words and sentences should be spaced.

space between sentences should be twice the space used between words. When several lines are required, the spacing between them may vary from one-half to one-and-one-half times the height of the letters.

Height of Letters and Numerals. The actual height of letters used on a drawing depends on the function of the composition. Titles and part numbers always have the greatest height, while letters for such items as section designation, part name, notes, dimensions, bill of materials, etc., are correspondingly smaller, See Fig. 6. Whole numbers and decimals should be made the same height as the letters. The height of each number in a fraction should be approximately three-fourths

	HEIGHT FREE HAND	EXAMPLE
Part Number in Title Block	1/4 to 3/8 3/8 Preferred	**3568529**
Section and Tabulation Letters	1/4	**A-A**
Part Name in Title Block	5/32	**CRANKSHAFT**
Revision Column, Process and Dimensional Notes	1/8	**FINISH ALL OVER**
Dimensions: Fractional, Decimal	1/8	$\frac{1}{8}$ $\frac{3}{4}$ .125 $\frac{1}{16}$
Revision Letter or Revision Number on Body of Drawing	1/8	Ⓐ ⑥
Sub Titles for Special Views	1/8	**ENLARGED VIEW**

Fig. 6. Recommended standard for height of letters.

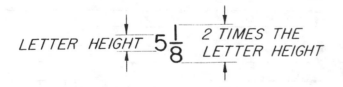

Fig. 7. This is how fractional numbers should be made.

the height of the whole number. The full height of the fraction including the fraction bar and space above and below the bar should be approximately twice that of the whole number. See Fig. 7.

VERTICAL	INCLINED	①
A	*A*	
B	*B*	
C	*C*	
D	*D*	
E	*E*	
F	*F*	
G	*G*	
H	*H*	
I	*I*	
J	*J*	
K	*K*	
L	*L*	
M	*M*	
N	*N*	
O	*O*	
P	*P*	
Q	*Q*	
R	*R*	
S	*S*	
T	*T*	
U	*U*	
V	*V*	
W	*W*	
X	*X*	
Y	*Y*	
Z	*Z*	

PROBLEM 2. *Letter pledge of allegiance in vertical letters.*

VERTICAL
(2)

I PLEDGE ALLEGIANCE TO THE FLAG OF THE UNITED

INCLINED
(3)

I WAS BORN ON

PROBLEM 3. *Letter brief autobiography of yourself in inclined letters.*

PROBLEM 4. *Complete each line of numbers.*

VERTICAL	INCLINED	④
1	*1*	
2	*2*	
3	*3*	
4	*4*	
5	*5*	
6	*6*	
7	*7*	
8	*8*	
9	*9*	
0	*0*	

VERTICAL   INCLINED   ⑤

$\frac{1}{32}$ = .031   *1 MILE =*   *FT.*

$\frac{1}{16}$ = .   *1 TON =*   *LBS.*

$\frac{1}{8}$ = .   *1 SQUARE MILE =*   *ACRES*

$\frac{3}{16}$ = .   *1 ROD =*   *FEET*

$\frac{1}{4}$ = .   *1 CUBIC FOOT =*   *CUBIC INCHES*

— = .375   *π = — OR*

— = .500   *$\frac{1}{4}$ π = .*

$\frac{5}{8}$ = .   *1 GROSS =*   *DOZEN OR*

— = .75   *1 REAM =*   *SHEETS*

$\frac{7}{8}$ = .   *1 INCH = — OF A FOOT*

$\frac{15}{16}$ = .   *1 BUSHEL =*   *PECKS*

PROBLEM 5. *Letter in the missing quantities.*

VERTICAL OR INCLINED

STATE

CAPITAL

NEW YORK   ALBANY

COUNTRY

CAPITAL

UNITED STATES   WASHINGTON D.C.

PROBLEM 7. *Letter the correct measurements.*

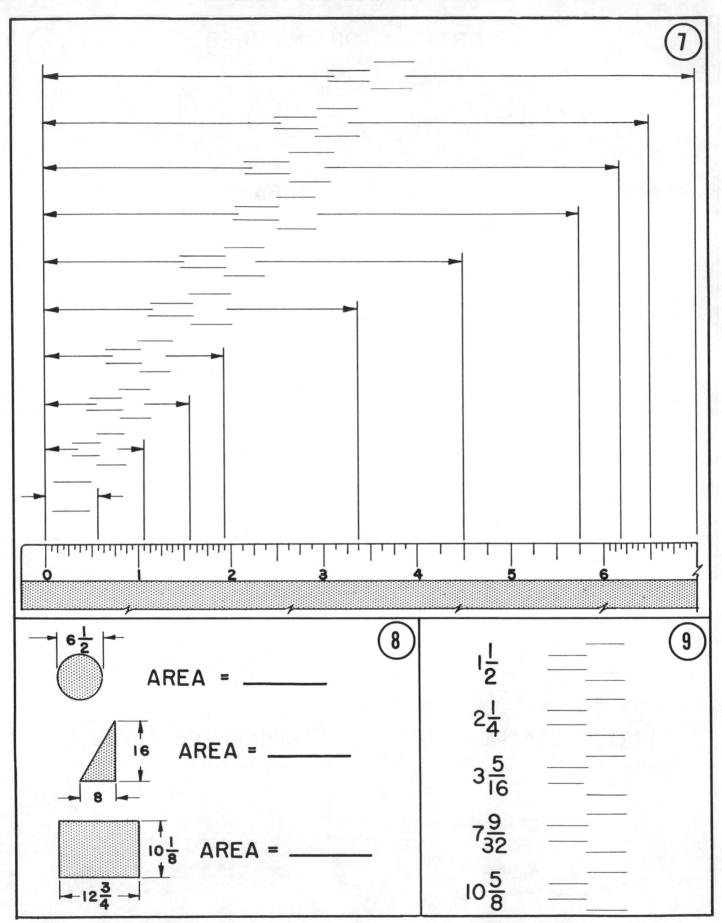

PROBLEM 8. *Insert answers to problems shown.*

PROBLEM 9. *Repeat the numbers given.*

The ability to produce a freehand sketch is an important asset to anyone associated with industrial and engineering work. Most ideas are usually expressed first through the medium of a freehand sketch and later translated into finished engineering drawings. Thus, the engineer or designer will very likely sketch out the preliminary features of a product and then pass them on to the draftsman for detailing. The draftsman frequently resorts to a freehand sketch to convey information of a mechanical drawing to the shop foreman. Even the foreman may be required to use some kind of sketch to show his workers a detail involved in the fabrication process.

THE

(10)

MAKING MULTIVIEW SKETCHES

A multiview drawing or sketch is often called a working drawing. It is a type of drawing which shows exactly how an object is constructed.

The accepted method of reproducing the accurate shape of an object is based on a system known as orthographic projection. In such a system a series of separate views are arranged so that each view is definitely related to the others. The material in this unit will tell you how to make multiview sketches.

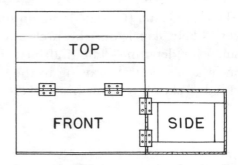

Fig. 2. Position of the views when in a flat plane.

hinged. When the sides are opened, each view falls in a certain position (Fig. 2). Thus you will see that the top view is directly above the front view, and the side view is opposite the front view. These, then, are the principal views of a working drawing, and these are the positions in which the views must be placed. The only difference is that in a working drawing the views are not shown in a connected position. Instead, they are drawn with a space between them, as in Fig. 3.

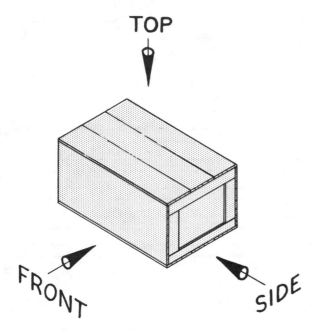

Fig. 1. These are the main views of a working drawing.

Type and Placement of Views. A working drawing has two or more views. The number of views that are shown depends on the shape of the object. To understand the relationship of the various views in a working drawing, look at the box shown in Fig. 1. Notice how these views appear as you look at the box from different positions.

Now assume that the edges of this box are

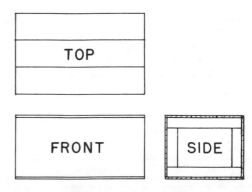

Fig. 3. In a working drawing, the views are spaced apart.

As a rule, for most working drawings, only the front, top, and side views are shown. It will be found, however, that some objects require only two views. This would be true in the case where the third view is simply a dupli-

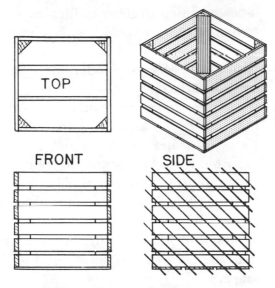

Fig. 4. In this drawing only two views are needed.

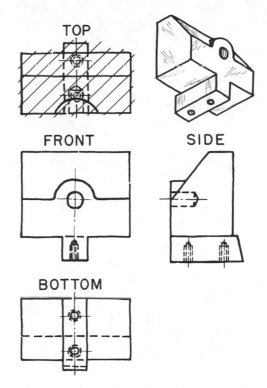

Fig. 6. This object needs a bottom view.

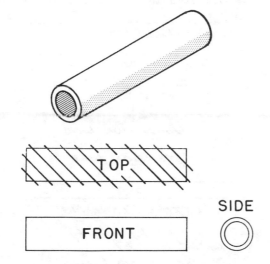

Fig. 5. A working drawing for this pipe needs only two views.

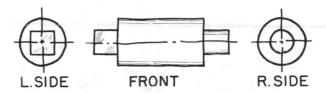

Fig. 7. This object needs a right and left side view.

cation of one of the other views. As an example of this, the square crate shown in Fig. 4. or the pipe illustrated in Fig. 5, obviously need only two views.

Occasionally there will be some objects which will require a bottom view as well as a top view. The bottom view is used only if the details cannot be clearly shown in the top view (Fig. 6).

The general practice is to use the right side view of the object. It may be that the shape of

the two ends differs to such an extent that their true shape cannot clearly be shown by a single side view. In such cases, the left side view must also be shown (Fig. 7).

Hidden Surfaces. A surface that cannot be seen in a view is shown by a hidden line. A hidden line, as you will see in Fig. 10, consists of short dashes. If you examine Fig. 8, you will see how hidden surfaces are indicated. When sketching hidden lines, always start with the dash in contact with the solid line, as shown in 1 of Fig. 9. A hidden line that runs into another hidden line should be drawn so the last dash touches the dash of the other, as in 2. When the hidden line is a continuance of a solid line, a space should be provided, as illus-

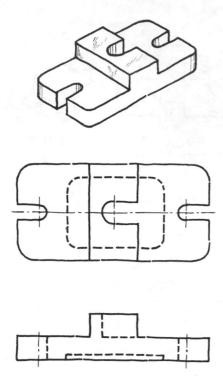

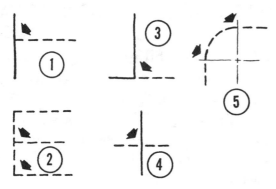

Fig. 8. Notice how hidden lines are used to show surfaces which cannot be seen.

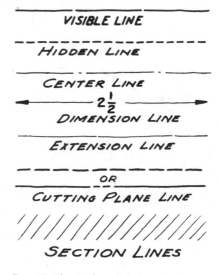

Fig. 9. These are examples of how hidden lines are drawn.

trated in 3. If the hidden line crosses a solid line, a space should be left, as in 4. Notice in 5 that dashes for hidden arcs start and end with a dash at points of tangency.

Alphabet of Lines. In sketching a working drawing, the following kinds of lines are used (Fig. 10):

1. *Visible lines* are heavy lines used to represent the visible edges of the object.
2. *Hidden lines* consist of short dashes approximately ⅛″ long with 1/32″ spaces.

They are used to show hidden surfaces.

3. *Center lines* are made with alternate long and short dashes. They indicate centers of objects, circles, and other curved surfaces.
4. *Dimension lines* are thin lines used to indicate the distance measured.
5. *Extension lines* run from the edges of the object, and indicate the points at which dimension lines are placed.
6. *Cutting plane lines* are made with equal dashes or alternating long and short dashes, and are used to indicate parts that are cut away to show the interior shape.
7. *Section lines* are thin slanted lines drawn across the cut surfaces of sections.

VISIBLE LINE

HIDDEN LINE

CENTER LINE

2½

DIMENSION LINE

EXTENSION LINE

OR

CUTTING PLANE LINE

SECTION LINES

Fig. 10. This is the alphabet of lines.

Weight of Lines. The various lines shown in Fig. 10 have three different weights—heavy, medium, and light. The contrast in weight should not be made by degrees of darkness but by differences in thickness. This can be done by varying the degree of pressure on the pencil.

In freehand sketching, it is a good practice to make all lines as light as possible to start with. The desired weight of lines can be attended to after the sketch is completed. In this way, lines can be easily corrected without having to erase a great deal.

Orientation of the Object. The object to be drawn can be turned so its principal surface is parallel to any plane of projection. For example, the object in Fig. 11 may be orientated so the surface *A* is parallel either to the vertical, horizontal, or profile plane. In each case it will be seen that the outline of the resulting views as they would normally appear on the drawing will vary depending on the orientation of the object.

The general rule for object orientation is to place the object so the sides having the most descriptive features are perpendicular to the direction of sight lines and parallel to a plane of projection. Specifically it is important to situate the object so the least number of hidden lines will appear in the views.

Spacing of Views. The views of an object should be arranged so as to present a balanced appearance on the drawing sheet. Ample space must be provided between views to permit the placement of dimensions without crowding, and to preclude the possibility of notes pertaining to a view from overlapping or crowding the other views. See Fig. 12.

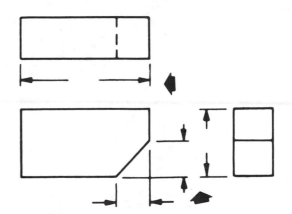

Fig. 12. Views should be balanced on the sheet with ample space between them.

Precedence of lines. In laying out a view there will be instances when hidden lines, visible lines, and center lines will coincide. Since the essential features of the object are important, visible lines must always take precedence over hidden lines or any other line. A hidden line always takes precedence over a center line. If a center line coincides with a cutting plane line, the line that contributes more to the readability of the drawing takes precedence. Dimension and extension lines should be placed so they will not coincide with other lines of the drawing. See Fig. 13.

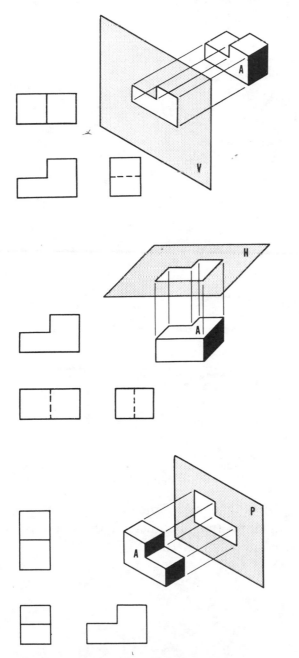

Fig. 11. The shape of the views will depend on the orientation of the object.

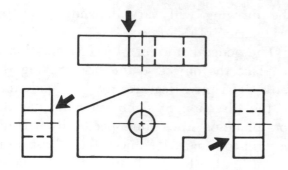

Fig. 13. When different types of lines coincide, the accepted precedence of lines must be observed.

Projection of Curved Surfaces. Many objects have curved surfaces or curved edges. When a circle or curve is parallel to the plane of projection, it is shown as a circle or curve on the parallel plane, and as straight lines on the adjacent planes. See Fig. 14.

If an object has continuously-curved surfaces that are perpendicularly tangent to each other, their line of tangency will be perpendicular to the plane of projection as shown in Fig. 15.

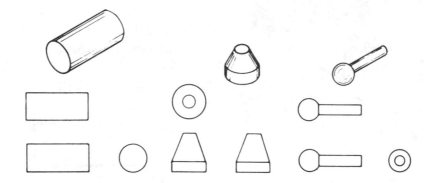

Fig. 14. This is how a curved surface appears on the various planes of projection.

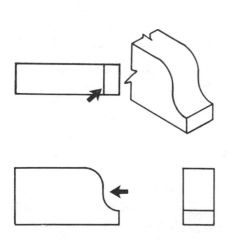

Fig. 15. The line of tangency of these two curves appears as a straight line in the top view.

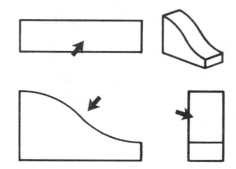

Fig. 16. If tangent curves are at an angle, no line is shown in the plane of projection.

If the tangent plane of two curves is at an angle, no line is shown in the plane of projection. See Fig. 16.

PRACTICE SHEET

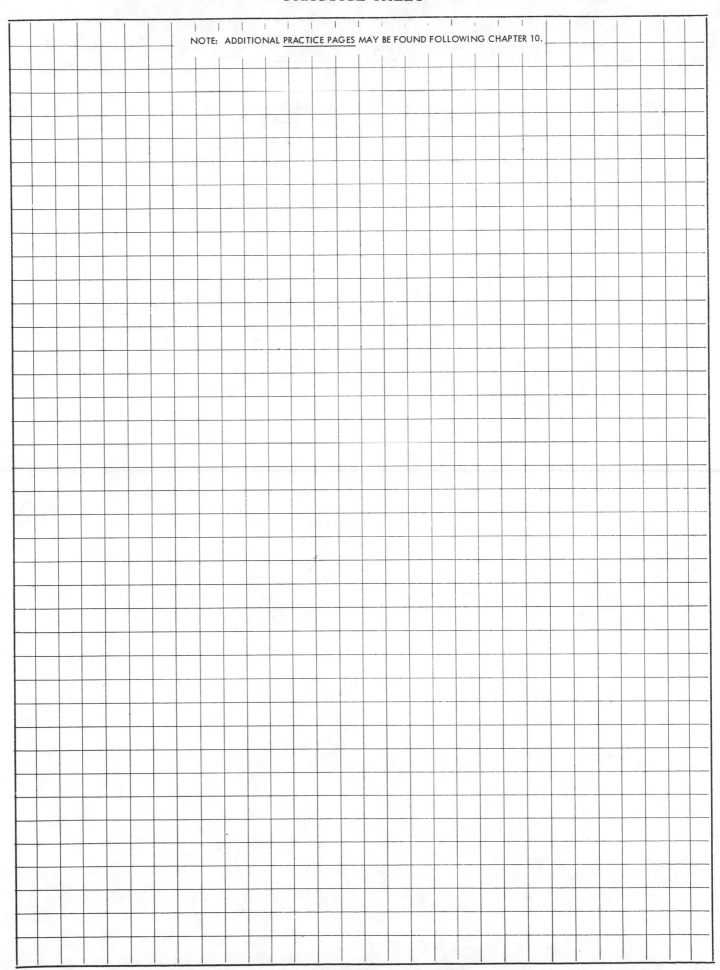

NOTE: ADDITIONAL PRACTICE PAGES MAY BE FOUND FOLLOWING CHAPTER 10.

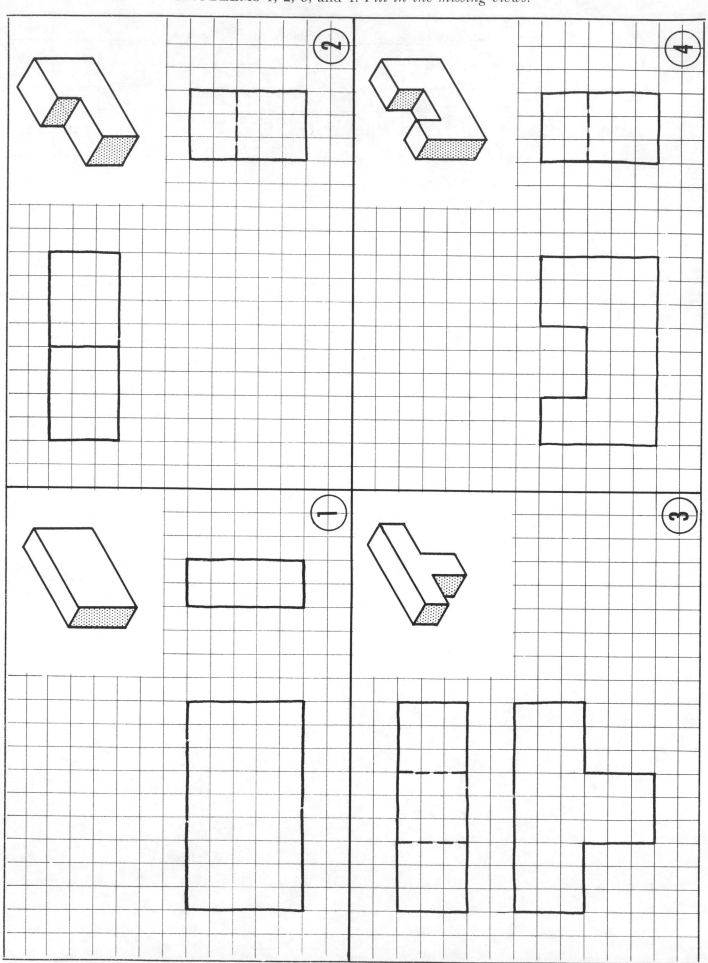

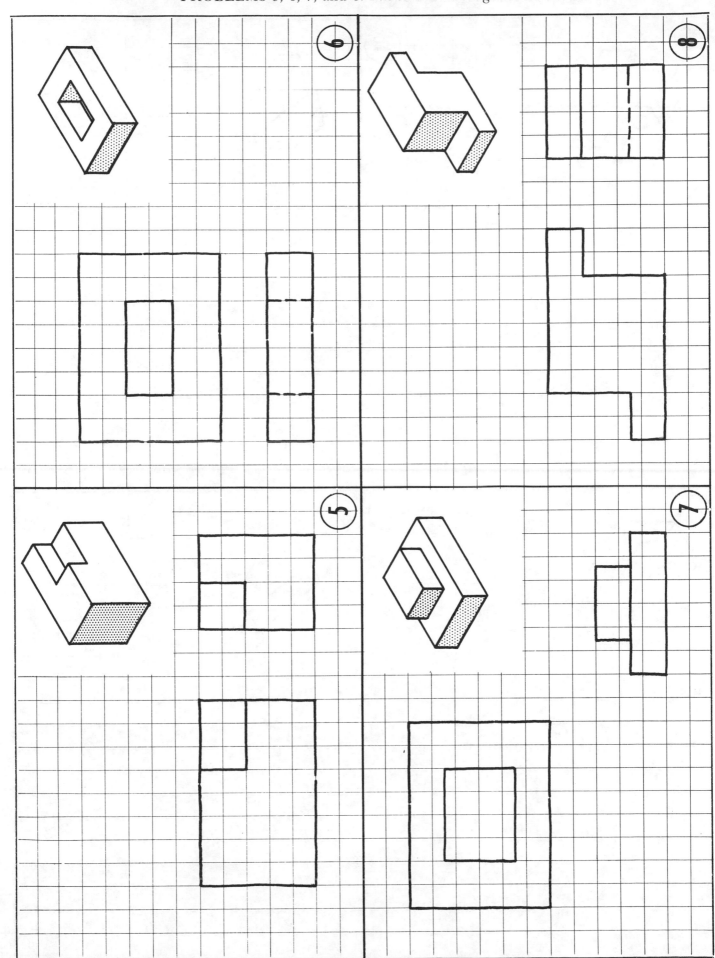

PROBLEMS 9, 10, 11, and 12. *Make a three-view sketch of the objects shown.*

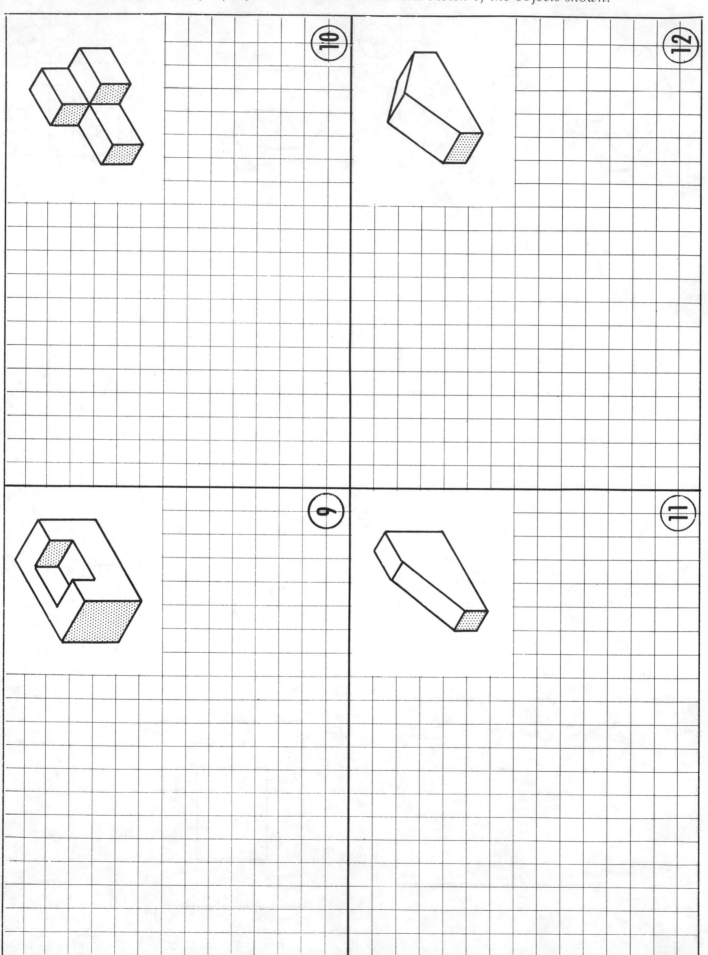

PROBLEMS 13, 14, 15, and 16. *Make a three-view sketch of the objects shown.*

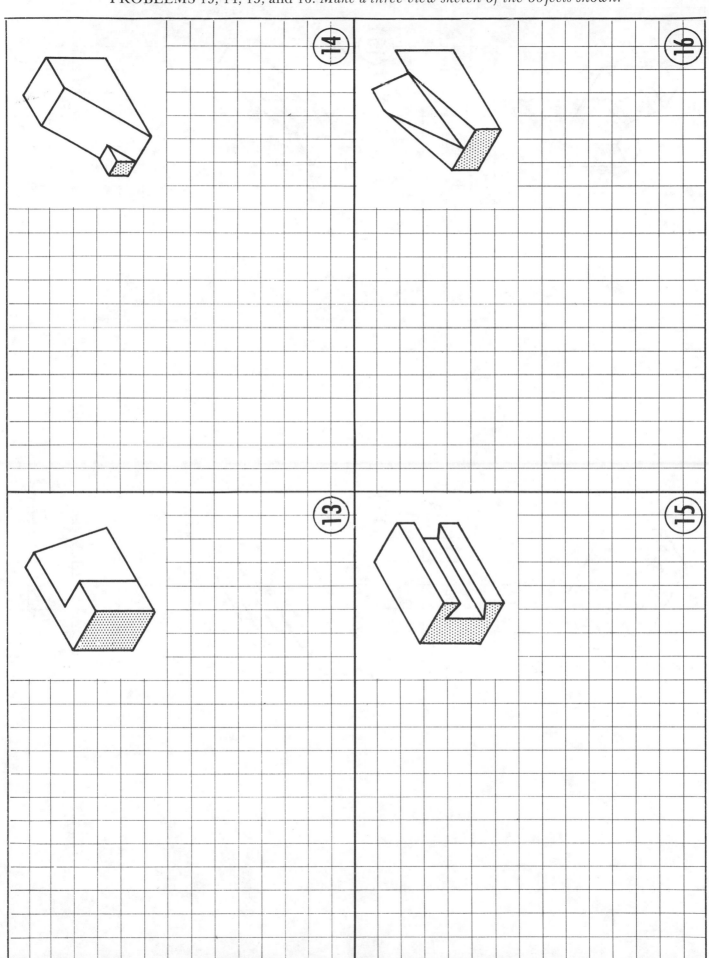

PROBLEMS 17, 18, 19, and 20. *Make a three-view sketch of the objects shown.*

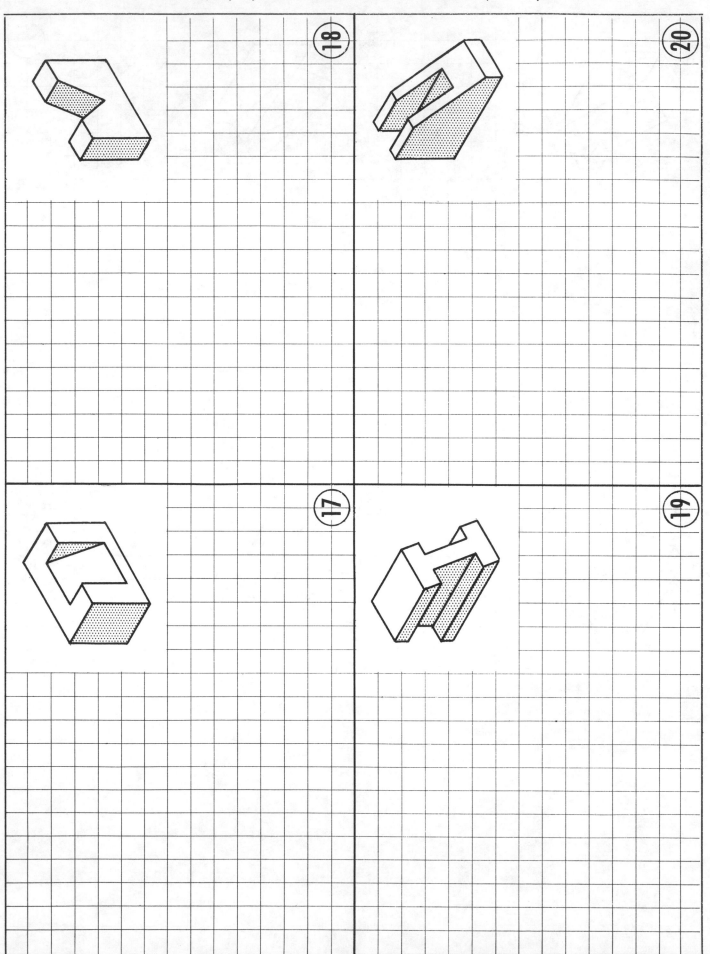

PROBLEMS 21, 22, 23, and 24 *Make a multiview sketch of the objects shown.*

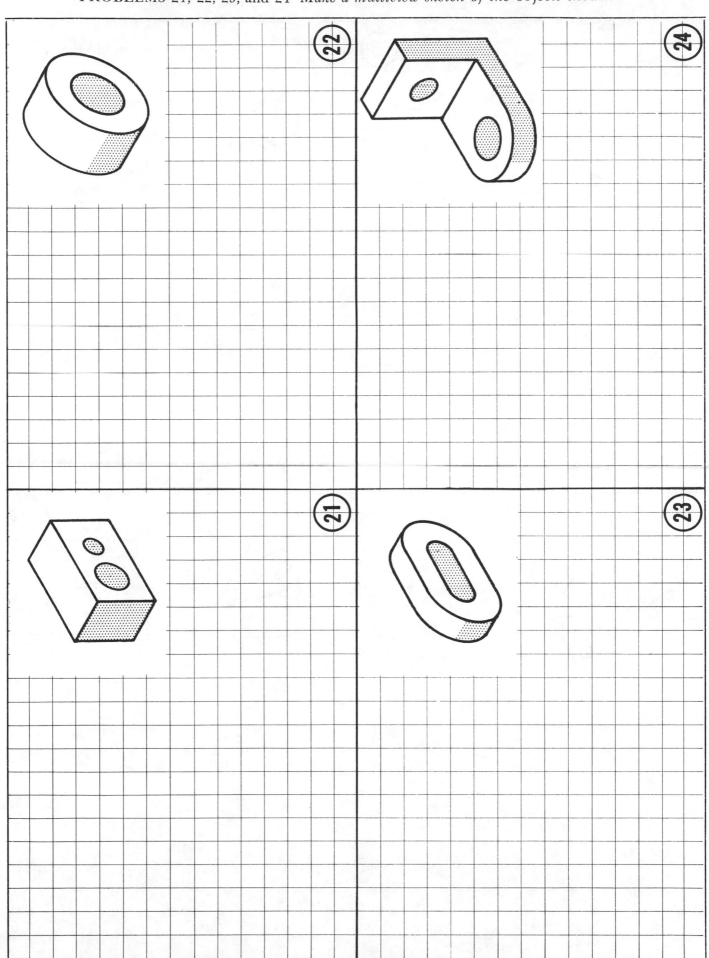

PROBLEMS 25, 26, 27, and 28. *Make a multiview sketch of the objects shown.*

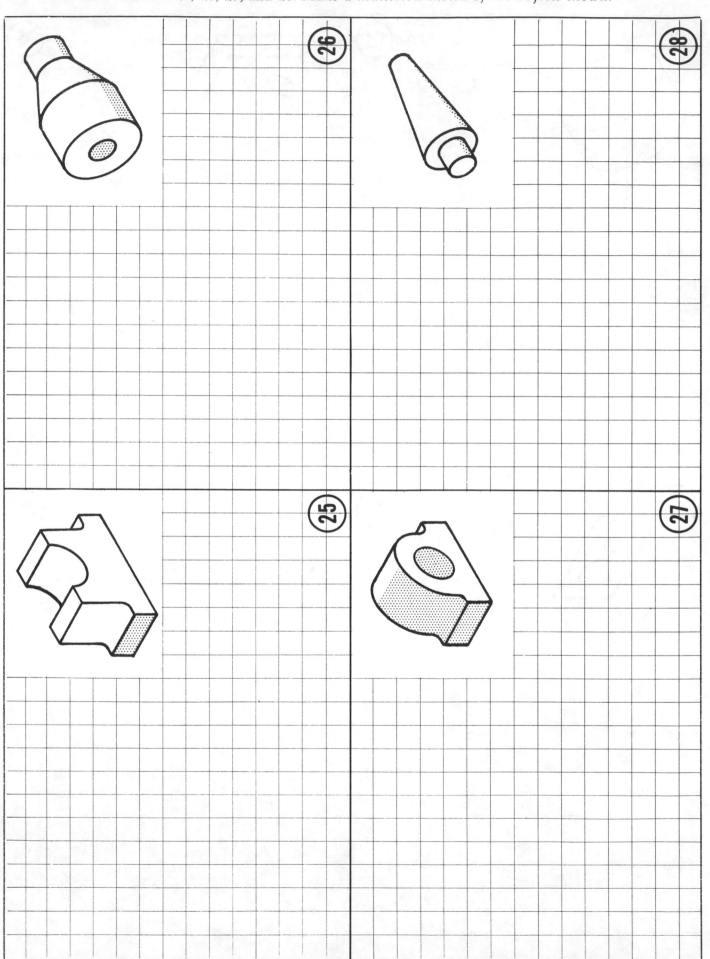

PROBLEMS 29, 30, 31, and 32. *Complete each view by filling in the missing lines.*

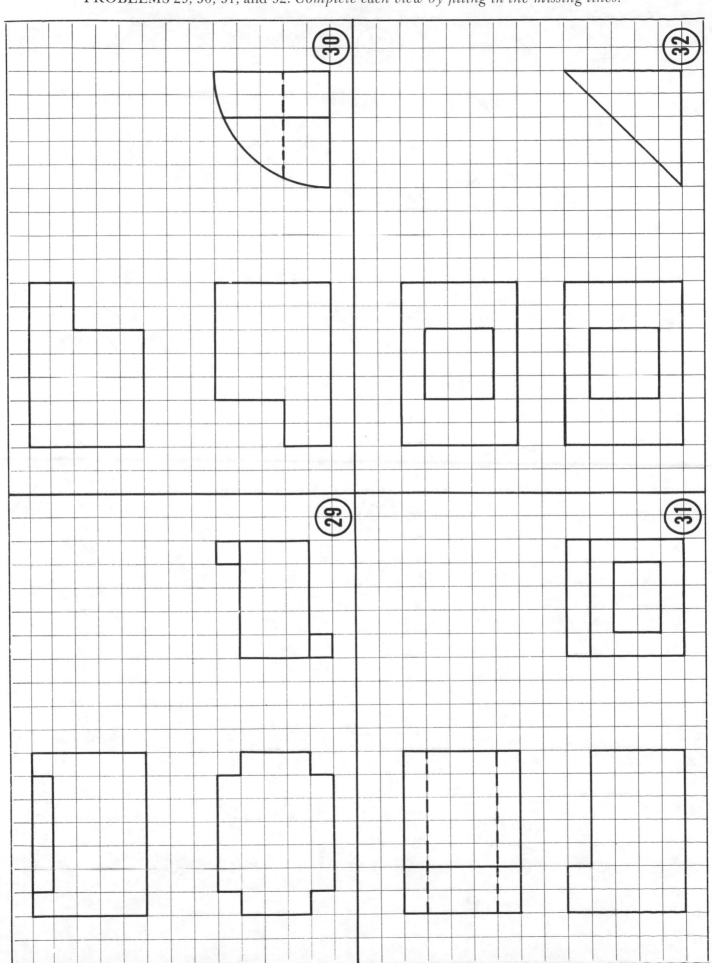

PROBLEMS 33, 34, 35, and 36. *Complete each view by filling in the missing lines.*

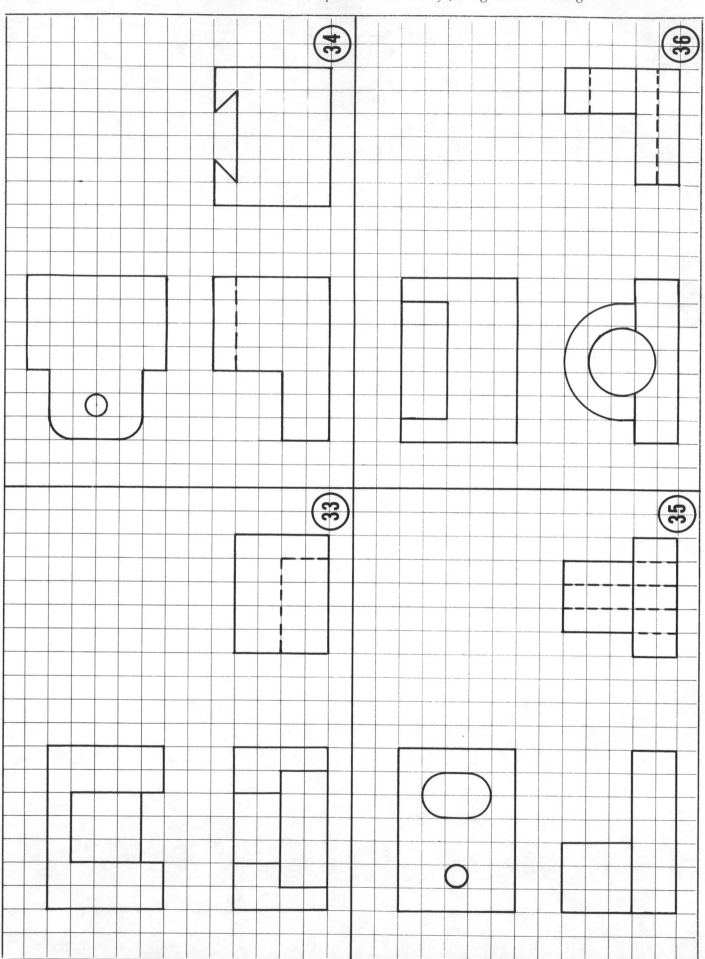

The value of a working drawing or sketch depends not only on the correct placement of the various views, but also on how accurately the dimensions are shown.

Dimensions may be expressed as common fractions or decimals. However, with the ever increasing importance of quality control in the manufacturing process, the decimal system of dimensioning is preferred.

In this unit you will have an opportunity to learn some of the basic dimensioning practices.

Units of Measurements. In the fractional dimensioning system the dimensions are given in inches and fractional parts of an inch, or in feet and inches. The symbol for the inch is ″ and for foot ′. The practice is to omit the inch symbol if all the dimensions are in inches. When dimensions are expressed in feet and inches, both the foot and inch symbols are used with a hyphen between them as 6′-4″, 5′-3½″.

The decimal system of dimensioning requires that all sizes be shown in decimal values. Ordinarily two- or three-place decimals are used depending on the degree of accuracy required. Generally speaking, in a two-place decimal, the second digit is usually rounded to an even number such as .02, .04, .08. (See decimal conversion chart inside back cover.)

How Dimensions Are Indicated. Angular and linear distances are shown by dimension lines with a break in the line for the insertion of the dimension figure. See Fig. 1. Dimension lines run between extension lines which should begin about 5/16″ away from the view and extend ⅛″ beyond the dimension line (Fig. 2).

Arrowheads, which are used on the ends of

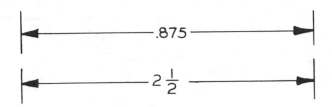

Fig. 1. Sizes are placed with dimension lines by these two methods.

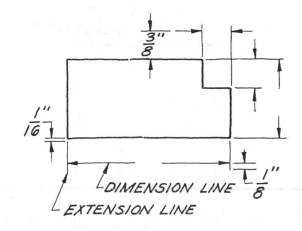

Fig. 2. Notice how extension lines are used.

dimension lines, should be approximately three times as long as they are wide. For most sketches they will average about ⅛″ in length. Check Fig. 3 and see how arrowheads are drawn. Try to keep all arrowheads uniform in size.

Angles should be dimensioned by an arc drawn with the vertex of the angle as a center

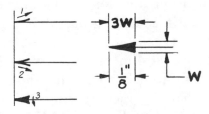

Fig. 3. Make arrowheads like this.

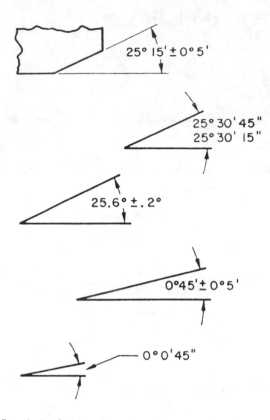

Fig. 4. Angles are dimensioned in a variety of ways.

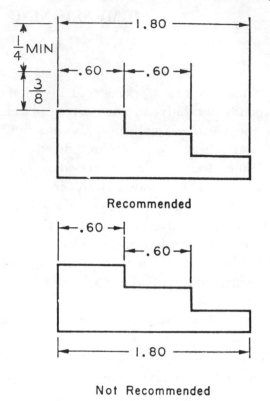

Fig. 5. Group dimension lines so they will produce an orderly appearance.

and the angular dimension inserted in a break in the arc. See Fig. 4. As a rule, right angles need not be dimensioned unless required for clarity. Angular dimensions should be expressed in ° (degrees), ' (minutes), and " (seconds). When degrees are used alone, the numerical value may be followed by the symbol °, or by the abbreviation DEG. If minutes are indicated alone, the value of the minutes should be preceded by 0°. It is permissible to dimension angles in degrees and decimal parts of a degree.

Dimension lines should be aligned whenever possible and grouped uniformly as shown in Fig. 5.

All parallel dimension lines should be not less than ¼ inch apart and should not be closer than ⅜ inch to the outline of the object. If several parallel dimension lines are necessary, the numerals should be staggered. Overall dimensions should be placed

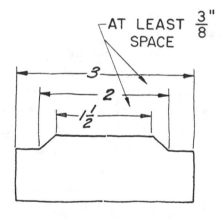

Fig. 6. Parallel dimension figures should be staggered.

outside the intermediate dimensions. See Fig. 6.

Extension lines should be drawn so they will not cross one another or cross dimension lines. Crossing lines can be kept to a minimum if the shortest dimension lines are drawn nearest the outline of the object. See Fig. 6.

General Dimension Arrangements. Here are some of the more basic rules for dimensioning a drawing:

1. Place dimensional figures so they read from the bottom of the sheet. See Fig. 7.

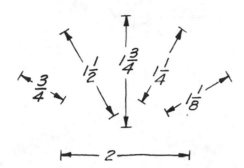

Fig. 7. Arrange dimension figures so they read from the bottom of the sheet.

2. Place dimensions outside of the part and, if possible, between views. See Fig. 8. Only in instances when the readability of a drawing is improved, should dimensions be placed within a view.

3. The most important dimension should be located at the principal view of a part because it is usually this view that most completely shows the essential contour characteristics of the piece.

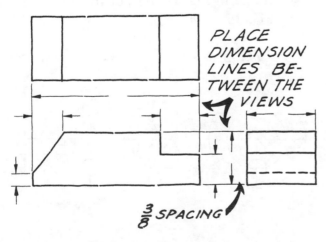

Fig. 8. Place dimension lines between views, at least ⅜" from the object.

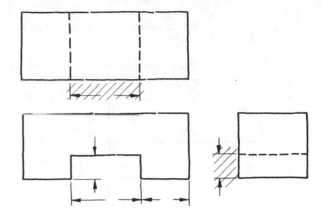

Fig. 9. Do not run dimensions to hidden lines.

4. Avoid placing dimensions which run to hidden lines (Fig. 9).

5. When dimensioning parts which are represented by broken lines, the dimension lines should remain unbroken (Fig. 10).

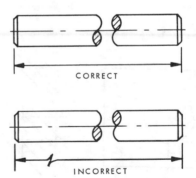

CORRECT

INCORRECT

Fig. 10. The dimension line should remain unbroken when dimensioning parts which are shown broken to conserve space.

6. Dimension a circle by giving the diameter. On small circles, place the dimension on the outside with the letters DIA. On large circles, place the dimension either on the inside or outside of the circle (Fig. 11). Always locate the dimensions of holes and circles on the view where they appear as circles, and *not* on the view where they are represented by hidden lines (Fig. 12). When

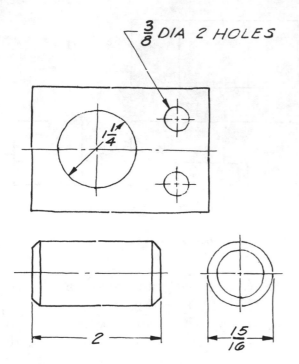

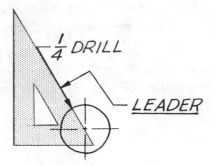

Fig. 13. To dimension the outside of a hole, draw a leader pointing toward the center of the circle.

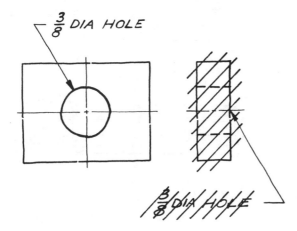

Fig. 11. Circles are dimensioned like this.

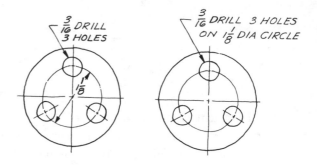

Fig. 14. This is how equally spaced holes are dimensioned.

Fig. 12. Do not dimension circles when they appear as hidden lines.

the circle diameter and number of holes, as in Fig. 14, *right*.

8. To dimension unequally spaced holes, show the angle of the holes, as illustrated in Fig. 15, *left*, or by offsets, as in Fig. 15, *right*.

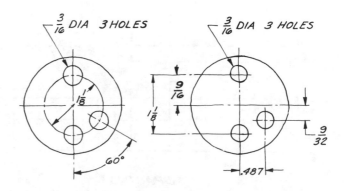

Fig. 15. This is how unequally spaced holes are dimensioned.

dimensioning the outside of holes, use a leader, as shown in Fig. 13.

7. When equally spaced holes are located around a circle, include the diameter of the circle across the circular center line. The number and size of the holes should be included in a note (Fig. 14, *left*). An alternate way is to show both

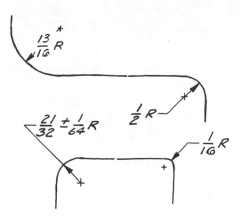

Fig. 16. Arcs are dimensioned like this.

9. Arcs should always be indicated by their radius, followed by the letter *R* (Fig. 16).

10. Place angular dimensions to read from the bottom of a drawing, except on large angles. On large angles, place the dimensions to read along the arc (Fig. 17).

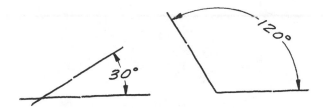

Fig. 17. Arrange angular dimensions in this manner.

11. Some features of objects are dimensioned from datums. Datums are lines, points, planes, or cylinders which are

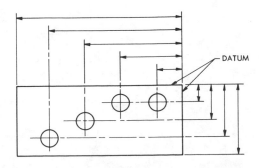

Fig. 18. Dimensions may be given from one or more datums.

assumed to be exact for purposes of computation or reference and from which the location of features of a part may be established (Fig. 18).

12. Symmetrical parts may be drawn fully dimensioned and with the addition of a note to specify "PART SYMMETRICAL ABOUT THIS CENTERLINE." On drawings of a symmetrical part where only a portion of the outline is pictured, due to the part's size or space limitations on the drawing, double arrows extending beyond the centerline are used to indicate symmetry. See Fig. 19.

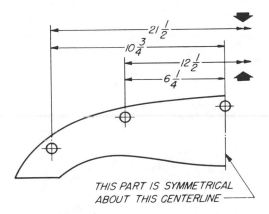

Fig. 19. How symmetrical parts may be dimensioned.

Limits and Tolerances. Limits are the maximum and minimum values prescribed for a specific dimension. A tolerance represents the total amount by which a specific dimension may vary, thus the tolerance is the difference between the limits.

A tolerance must be expressed in the same form as its dimension; the tolerance on a decimal dimension should be expressed by a decimal to the same number of places. For example, suppose a size is shown as 3.25". Variation from this size may be specified as being a plus .002" and a minus .003". Then the extreme permissible dimension could not

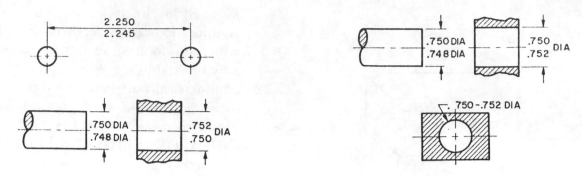

Fig. 20. Limit dimensioning.

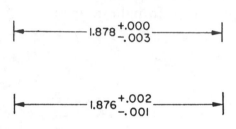

Fig. 21. Plus and minus tolerancing.

exceed 3.25″ plus .002 or 3.252″, or be less than 3.25″ minus .003″ or 3.247″.

Dimensional tolerances are expressed either as limit dimensioning or plus and minus tolerancing. In limit dimensioning the high limit is placed above the low limit as in Fig. 20 *left* or in note form as in Fig. 20 *right*. With the plus or minus system, the tolerance follows the specified size. See Fig. 21.

PROBLEMS 1 and 2. *Completely dimension each object shown.*

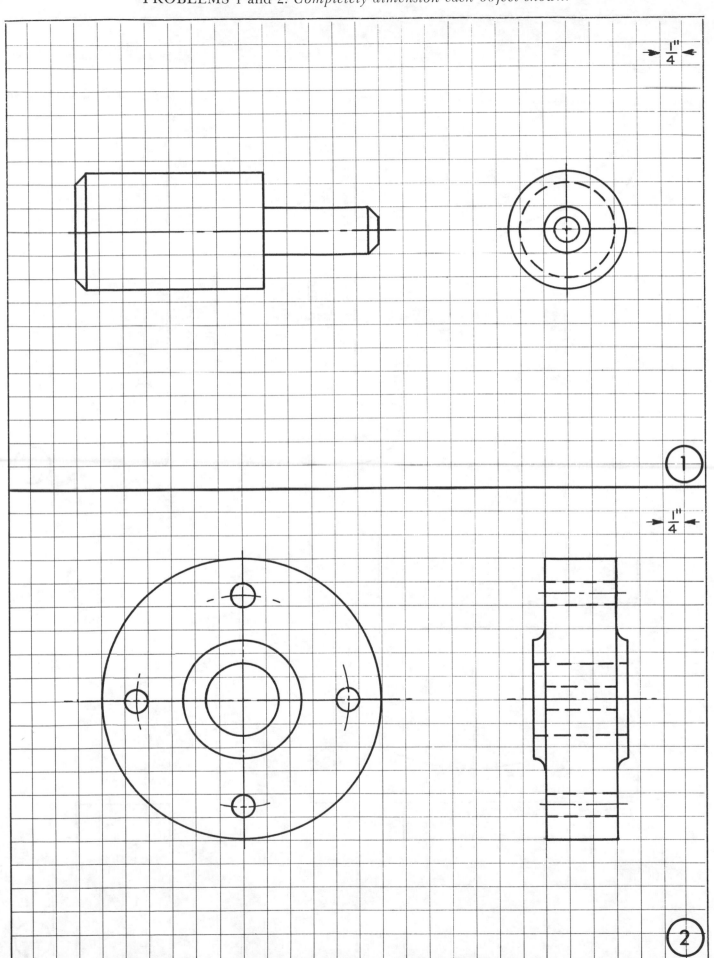

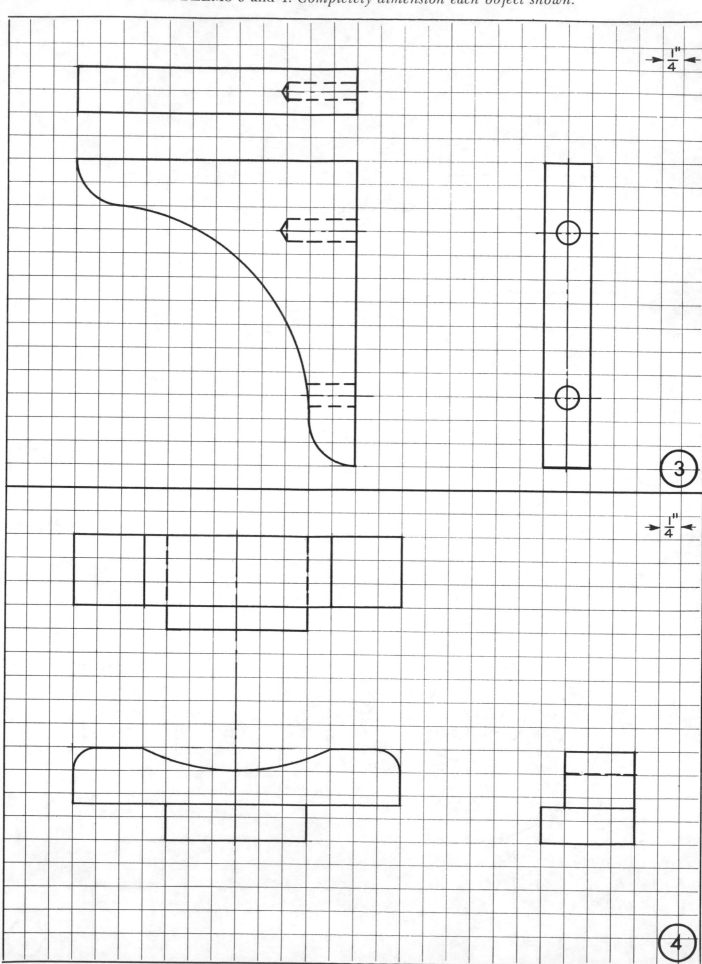

PROBLEMS 5 and 6. *Completely dimension each object shown.*

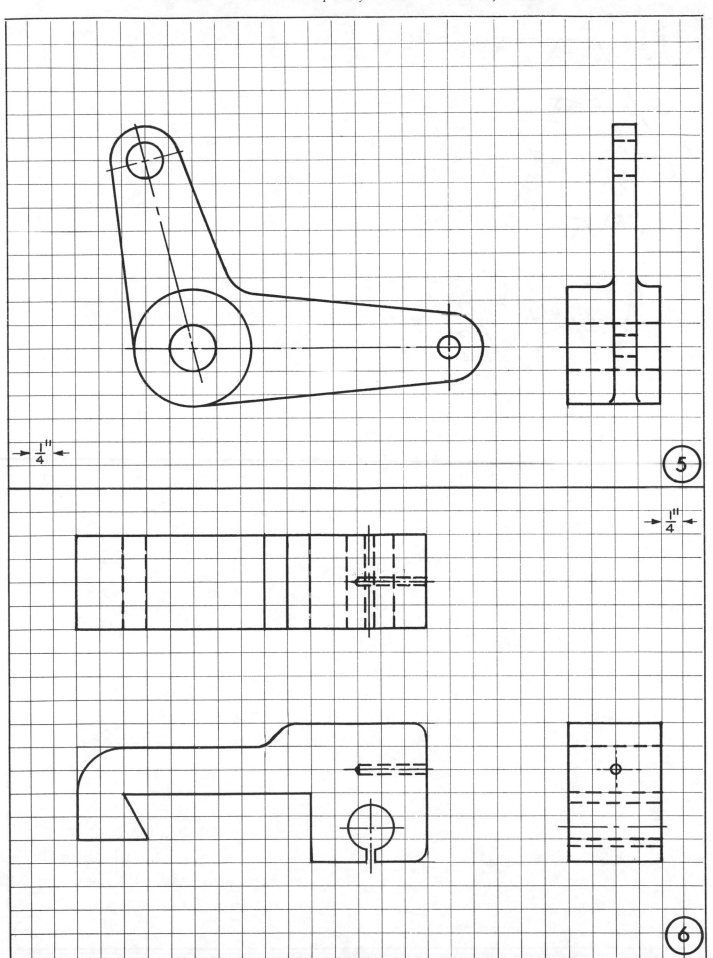

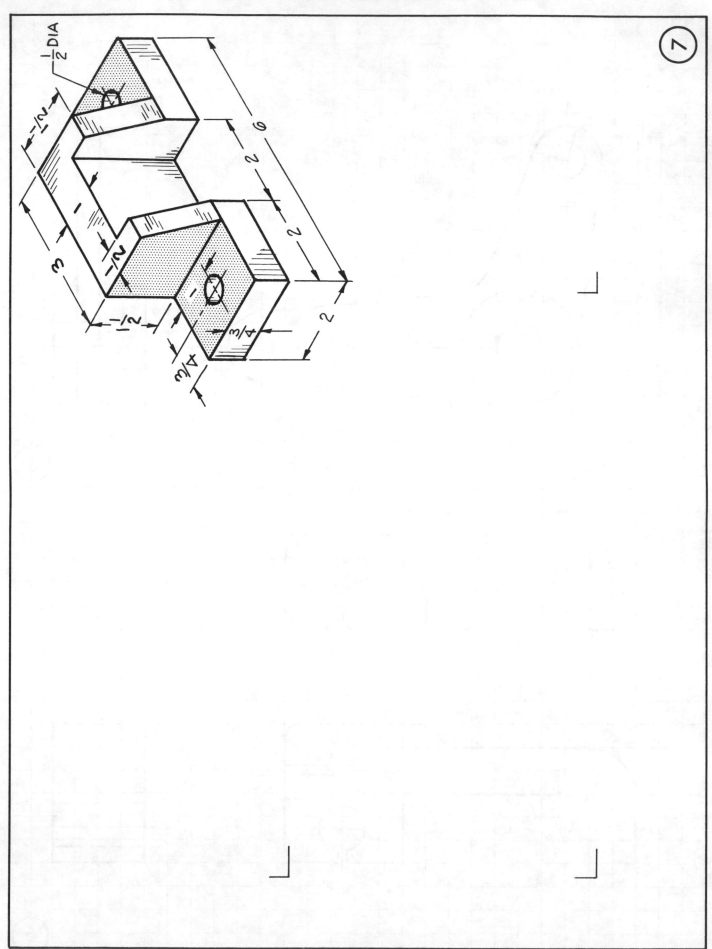

PROBLEM 8. *Make a multiview sketch and dimension. Convert all fractional sizes to decimal dimensions.*

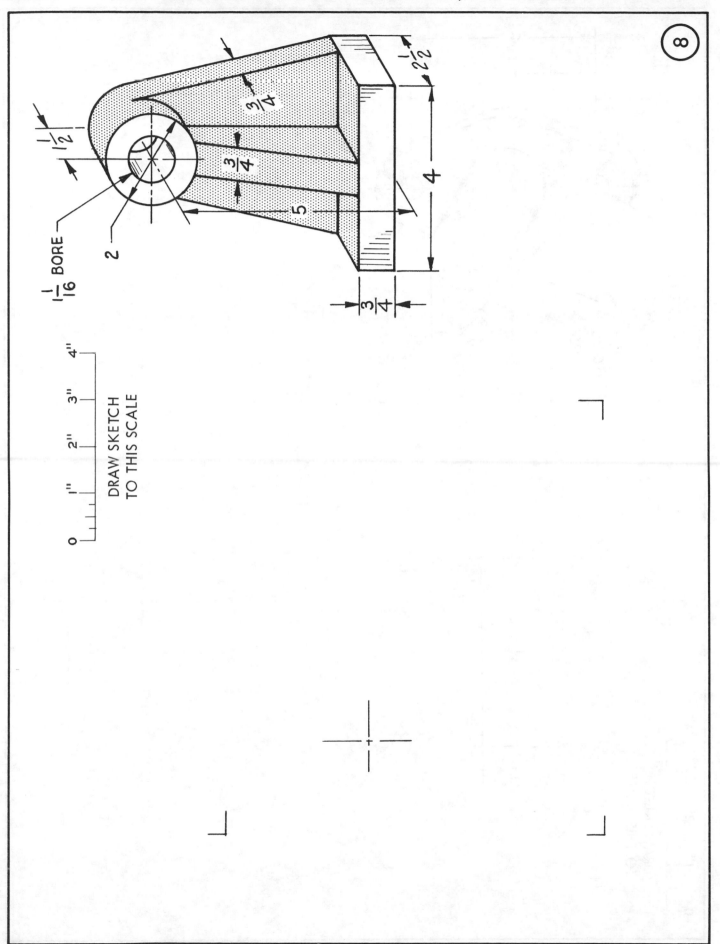

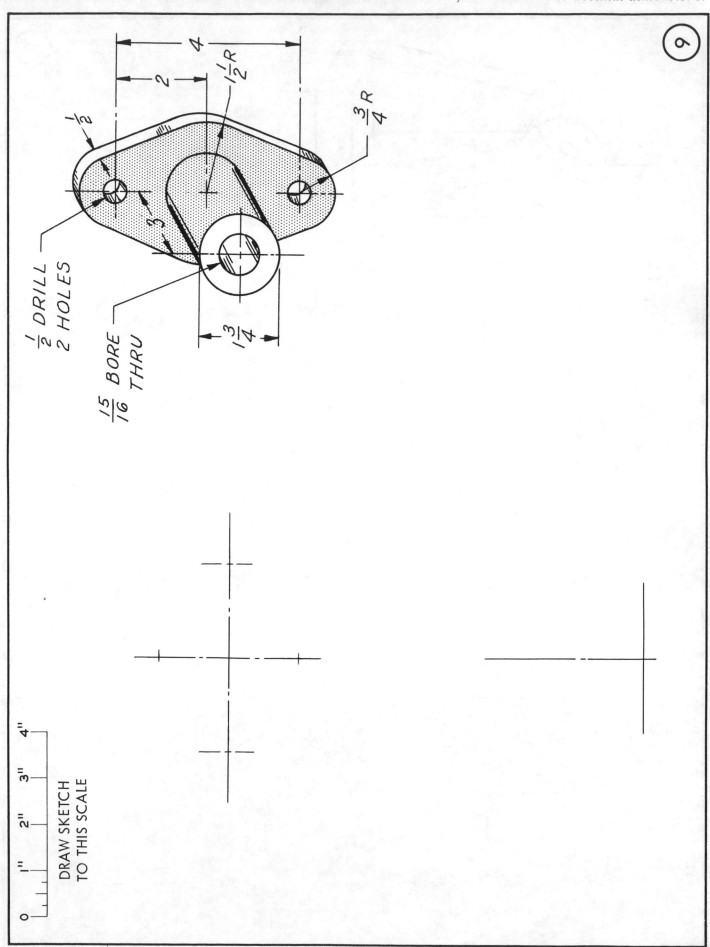

9

4

2

$\frac{1}{2}R$

$\frac{3}{4}R$

$\frac{1}{2}$

3

$\frac{1}{2}$ DRILL
2 HOLES

$\frac{15}{16}$ BORE
THRU

$\frac{3}{4}$

DRAW SKETCH
TO THIS SCALE

0   1"   2"   3"   4"

PROBLEM 10. *Make a multiview sketch and dimension. Convert all fractional sizes into decimals. Show all sizes with a .005" plus and minus tolerance using limit dimensions. Use any size sketch but maintain proper proportions.*

PROBLEM 11. *Make a multiview sketch and dimension. Convert all fractional sizes into decimals. Show all sizes with a .010" plus and minus tolerance. Use limits. Use any size sketch but maintain proper proportions.*

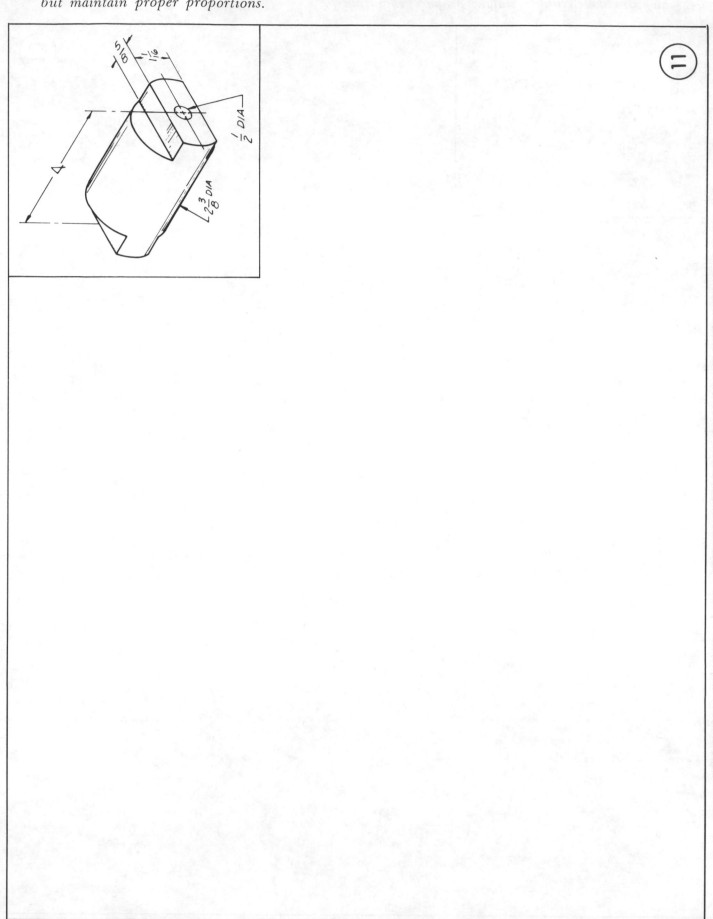

UNIT 5

SKETCHING SECTIONAL VIEWS

When making a multiview sketch, hidden surfaces and edges are usually indicated by hidden lines, as described in Unit 3. Sometimes the interior shape of an object is too complicated to show the construction by hidden lines. For a clearer presentation, the practice is to sketch the object with a portion of it removed. By removing a small section, the inside can easily be sketched without having to use numerous hidden lines on other views.

The cross-section of an object is obtained by passing an imaginary cutting plane through the object. The cutting plane is assumed to pass through at some selected portion of the object and the cut part is removed. See Fig. 1.

You will learn how sectioning is done by following the instructions in this unit.

Cutting Plane Line. The cutting plane is shown on the regular view by means of a cutting plane line. The cutting plane line is made either with alternating long dashes and pairs of short dashes or with equal dashes. The long lines should be about 3/4" to 1" and the short dashes 1/8" to 3/16" with 1/16" space between them. See Fig. 2.

The ends of both types of lines are bent

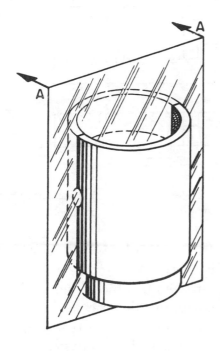

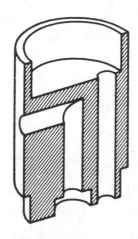

Fig. 1. By passing a cutting plane through the object, a portion can be removed to reveal the internal shape.

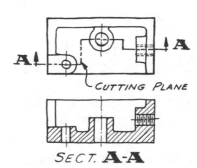

Fig. 2. This is how sections are labeled.

57

at 90° and terminate with arrowheads. The arrowheads should point in the direction of sight in which the object is viewed when the sectional view is made. Capital letters such as *A-A, B-B, C-C,* etc., are used to identify the section. A notation is also placed under the view and labeled *SECT. A-A, SECT. B-B,* etc. See Fig. 2.

Types of Sectional Views. If an object is cylindrical or symmetrical in shape, it is necessary to remove only one quarter of it. Such a section is known as *half section.* In this case, the imaginary cutting plane passes through one vertical center line and along another center line at right angles to the first (Fig. 3).

SECT. **A-A**

Fig. 4. This is how a full section appears when the cutting line passes along the main axis.

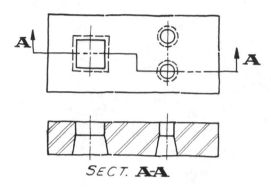

Fig. 5. Here is an example of a full section with the cutting line in an offset position.

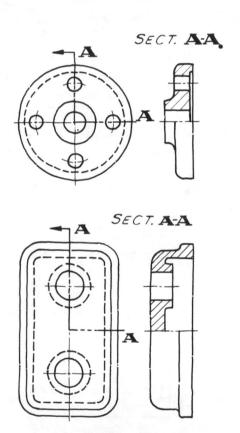

SECT. **A-A**

SECT. **A-A**

Fig. 3. These are half-sectional views.

A rectangular or square object, with an irregular interior, is shown by passing a cutting plane entirely through the object. The resulting section is referred to as a *full section* (Fig. 4). The imaginary cutting plane may either

pass along the main axis or it may be offset, as shown in Fig. 5.

For certain types of objects, such as bars, channels, spokes, or ribs, the sectional view must be *revolved* to obtain its true shape. In this case, the cutting plane is passed perpendicularly to the axis of the object and the section is turned through a 90° angle (Fig. 6). Frequently, the revolved section is pulled out from its position and located in some other convenient place on the paper; it is then called a *removed* section.

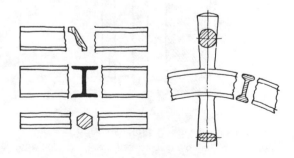

Fig. 6. This is how revolved sectional views are sketched.

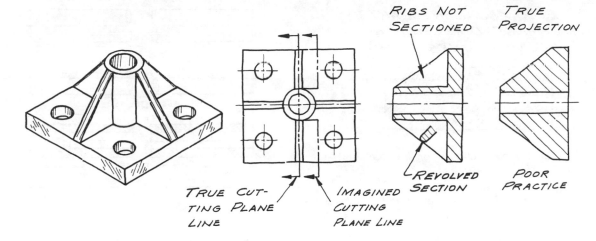

Fig. 7. Web and riblike parts are not lined with section lines.

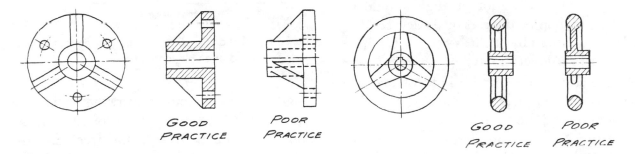

Fig. 8. Ribs and holes should be rotated in the plane of projection in order to present a symmetrical appearance.

Web and riblike parts of an object are not sectioned in a sectional view. Although the cutting plane line passes lengthwise through the center of the rib, the cutting plane should be imagined to be offset so it passes just in front of the ribs. The rib, then, is shown in outline form without section lines. Such a procedure is less likely to convey the impression that the object is solid. Usually, a part of the rib is revolved to show all that needs to be known about the shape of the rib (Fig. 7).

Where ribs and holes are arranged radially from a common center, they are rotated in the sectional view to convey a symmetrical appearance. By rotating them in this manner, it is easier to show their true distance from the center of the object (Fig. 8).

Representing Sectional Views. Whenever a sectional view is used, the surfaces of the sectional view are covered with thin parallel lines spaced approximately $\frac{1}{32}''$ or more apart and drawn at an angle of 45 degrees. Ordinarily, section lines should run in one direction, except where there are two or more adjacent pieces (Fig. 9). In such cases, the

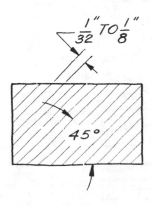

Fig. 9. Section lines should run in one direction, except where there are adjacent parts.

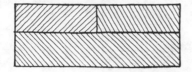

FOR ADJACENT PARTS
CHANGE DIRECTION OF
SECTION LINES

Fig. 10. Change the direction when adjacent parts are sectioned.

direction of the lines is changed so they are at right angles to each other (Fig. 10). If the shape or position of the part brings the 45-degree sectioning parallel, or nearly parallel, to one of the sides, another angle should be selected, such as 30 degrees or 60 degrees.

In sectioning objects, it is often a common practice to use certain types of section lines that actually represent the kind of material from which the piece is made. Fig. 11 illustrates the sectioning code approved by the American National Standards Institute.

Conventional Breaks. When long objects with small cross-sections and uniform in

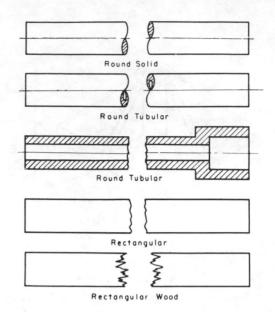

Fig. 12. Conventional breaks are used to shorten a view of an elongated object.

length do not fit onto a drawing sheet, the practice is to delete a portion of the object. Convention break lines made freehand are then used to indicate this condition. See Fig. 12.

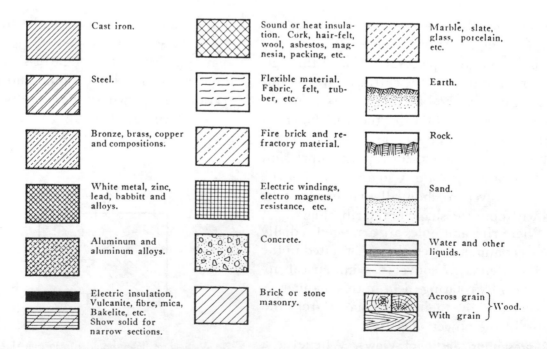

Fig. 11. Code symbols for section lines are shown here.

PROBLEM 1. *Complete the side view in full section.*

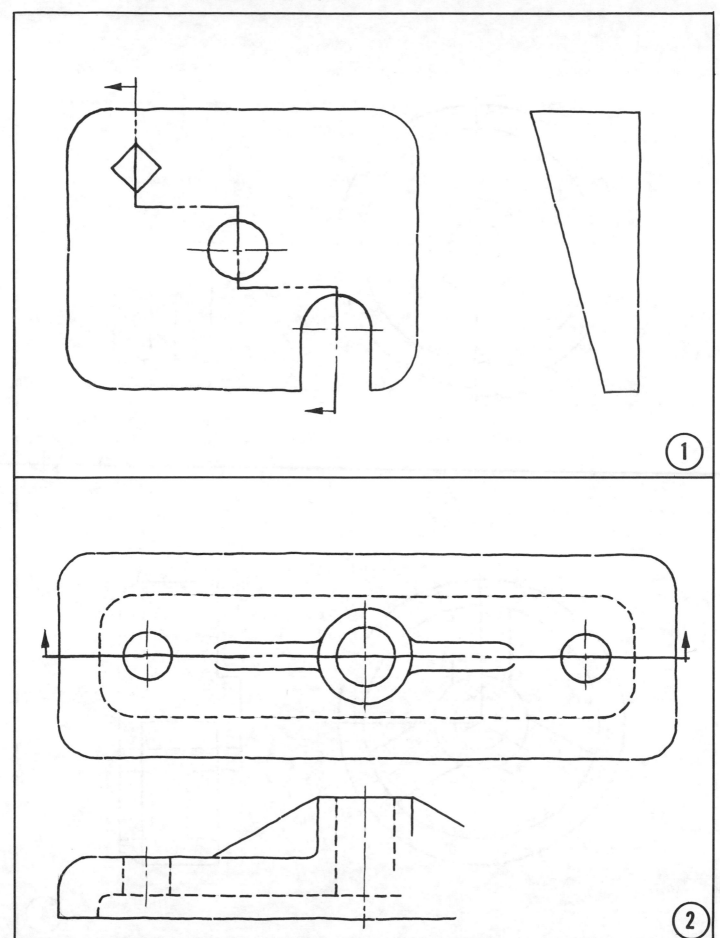

①

②

PROBLEM 2. *Complete the front view in full section; material cast iron.*

PROBLEM 3. *Complete the side view in full section; material steel.*

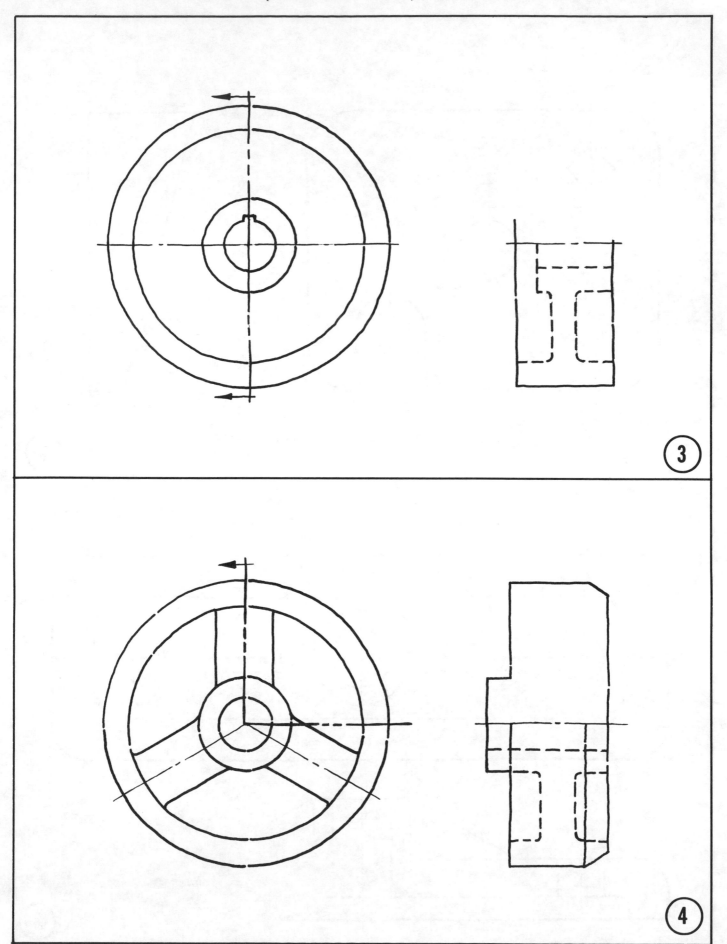

③

④

PROBLEM 4. *Show half-sectional side view, material brass.*

PROBLEM 5. *Complete the side view; material steel.*

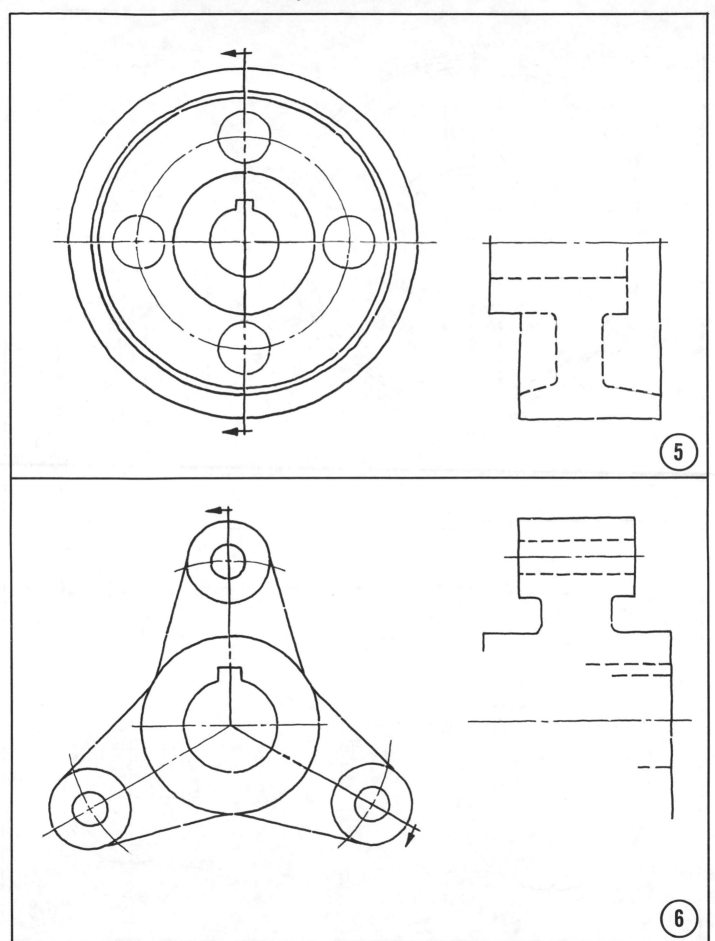

PROBLEM 6. *Section side view; material white metal.*

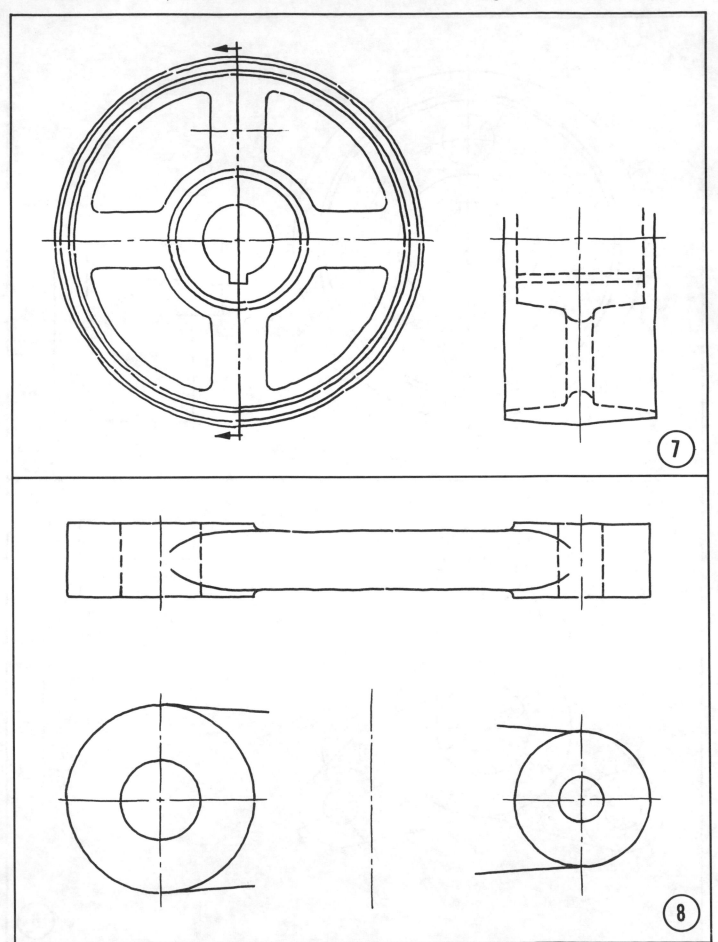

PROBLEM 8. *Show revolved section through elliptical arm; material steel.*

UNIT 6

SKETCHING AUXILIARY VIEWS

When an object has one or more inclined surfaces, as shown in Fig. 1, it is generally impossible to show the true shape of these inclined surfaces in the principal view of a working drawing. To present a more accurate description of any inclined surface, an addi-

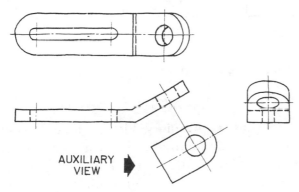

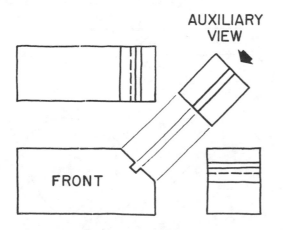

Fig. 1. To show the true shape of this object, an auxiliary view is required.

tional view, known as an auxiliary view, is required. The material in this unit describes how you can sketch auxiliary views.

Types of Auxiliary Views. There are three main types of auxiliary views. The first type is one in which the auxiliary view is projected from the front view (Fig. 2). The second type is one in which the auxiliary view is projected

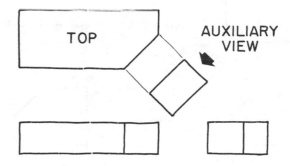

Fig. 3. This auxiliary view is projected from the top view.

from the top view, as illustrated in Fig. 3. In the third type, the auxiliary view is projected from the side view (Fig. 4).

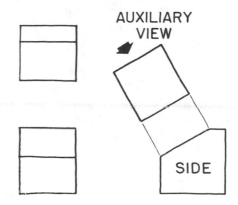

Fig. 4. This auxiliary view is projected from the side view.

Sketching Auxiliary Views. First, sketch the three principal views of the object, the front, top, and side views. Draw a reference line, AB, to represent the edge of the auxiliary view (Fig. 5). Then project the necessary points from the slanted surface to the reference line, and complete the auxiliary view. Whether the auxiliary view is to be projected from the front, top, or side view depends on the position of the object, or which surface of the object is slanted. Thus, in Fig. 5, the inclined surface is shown in the front view and therefore the auxiliary view must be projected from this view.

Fig. 2. This auxiliary view is projected from the front view.

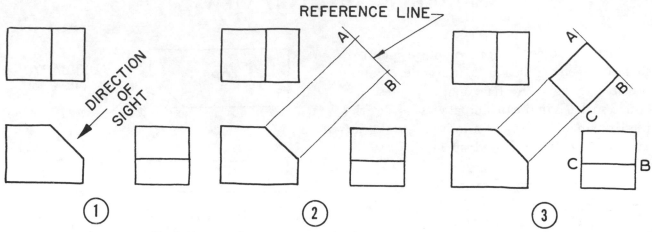

Fig. 5. Here are the main steps in sketching a simple auxiliary view.

When the inclined surface is symmetrical, as shown in Fig. 6, draw a center line, AB, for the auxiliary view parallel to the inclined face. Draw another center line, CD, through the top view. Project points EF, GH, and IJ from the top view to the slanted edge, and then to the auxiliary line. Take the necessary measurements from the top view and lay them to the right and left of the auxiliary center line. Then complete the view by joining these points.

Drawing an Auxiliary View with a Curved Surface. For an auxiliary view with a curve, as shown in Fig. 7, divide the curve or the top view into any convenient number of parts and project these points across to the slanted edge of the front view. From the front view, project the points to the center line of the auxiliary view. Then take each distance, shown as A, in the top view and lay it off on the auxiliary view. Continue with the remaining lengths and then sketch in the curve.

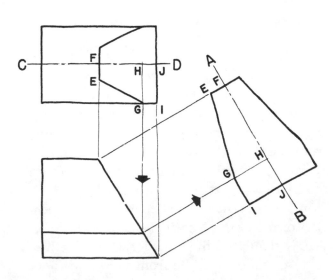

Fig. 6. This is an example of a symmetrical auxiliary view.

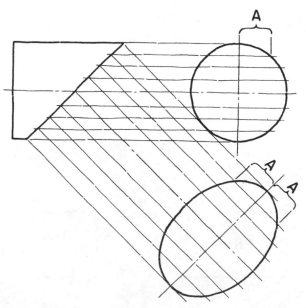

Fig. 7. This is how a curved line auxiliary view is drawn.

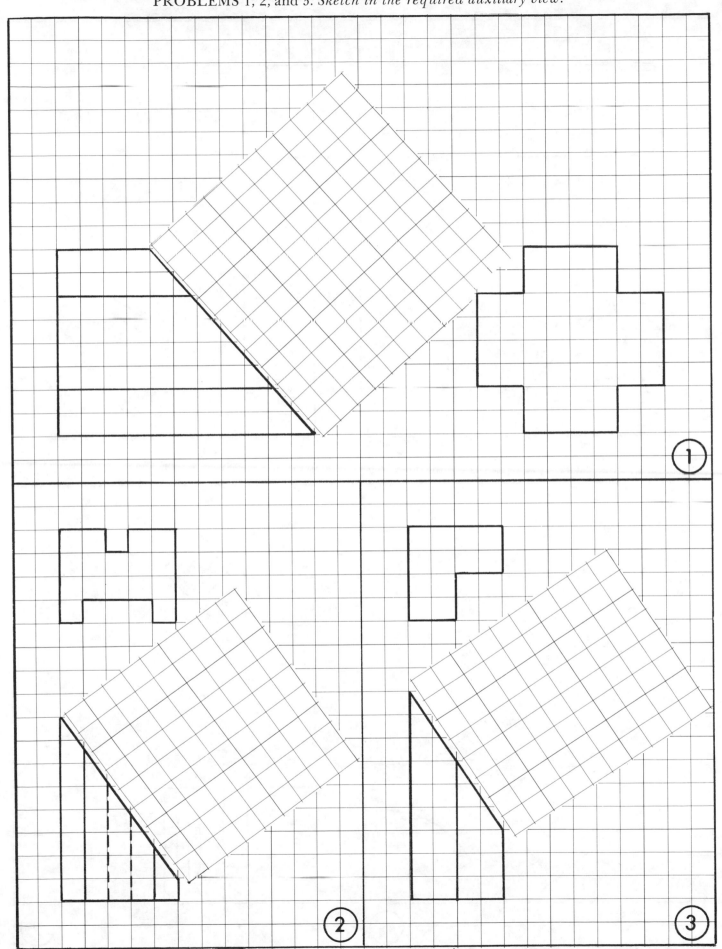

67

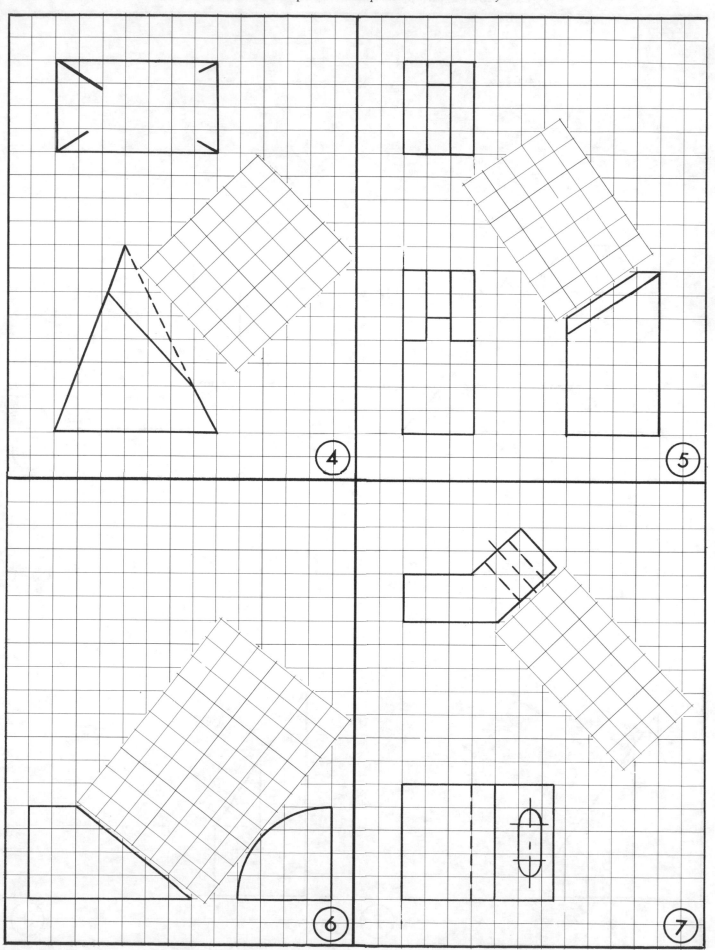

PROBLEMS 5, 6, and 7. *Sketch in the required auxiliary views.*

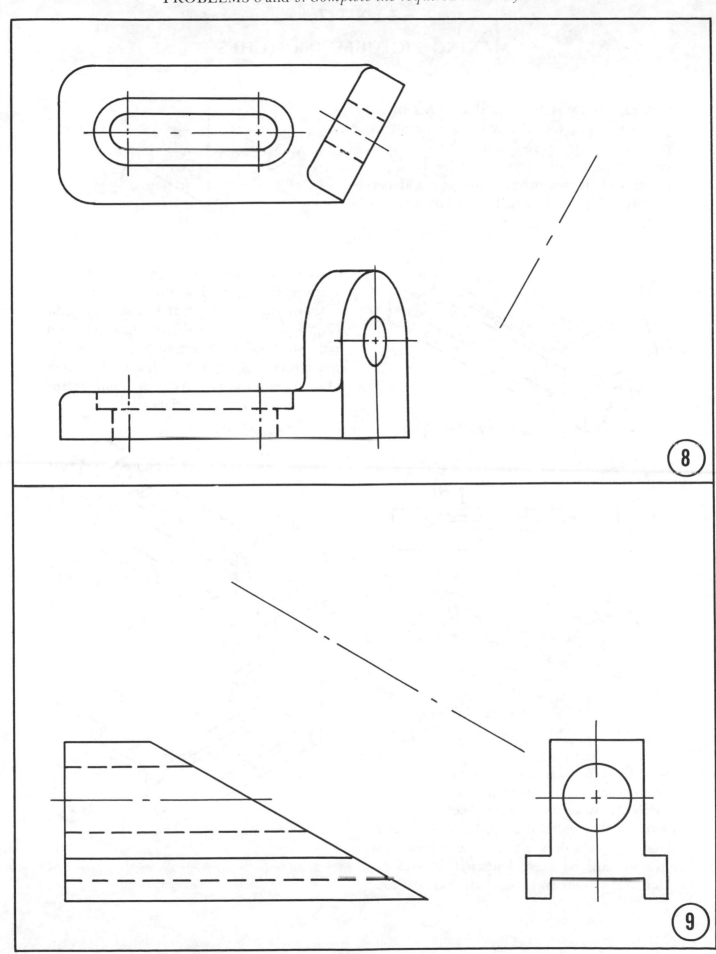

UNIT 7

MAKING PICTORIAL SKETCHES

A pictorial sketch is one that shows an object as it appears to the eye or as it would appear if a picture were taken of it with a camera. There are three principal kinds of pictorial sketches—isometric, oblique, and perspective (Fig. 1). You will learn how to make

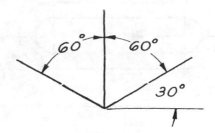

Fig. 2. This shows the main axis of an isometric sketch.

at an angle of 60 degrees (30 degrees to the horizontal) (Fig. 2). The object can be rotated so that either the right or left side is visible (Fig. 3). Whether the object is drawn with its main surfaces extending to the right or left depends entirely upon which side is the most advantageous to show. As a rule, hidden lines are never used on a pictorial sketch.

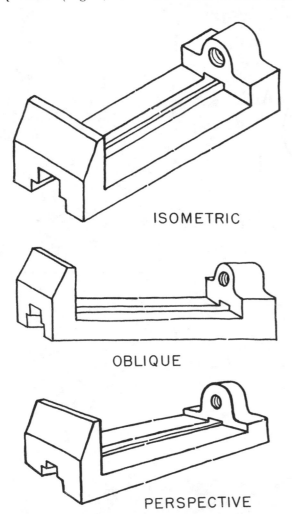

ISOMETRIC

OBLIQUE

PERSPECTIVE

Fig. 1. Here are several kinds of pictorial sketches.

these different pictorial sketches by following the instructions in this unit.

How To Make an Isometric Sketch. An isometric sketch shows three sides of an object. These sides are drawn along one vertical axis and two angular axes. The angular axes extend to the right and left of the vertical axis

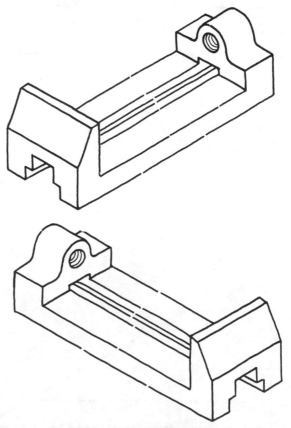

Fig. 3. An isometric sketch can be made with the object rotated to the right or left.

70

To make an isometric sketch, proceed as follows:

1. Draw a vertical line. From the top or base of this line, extend two slanted lines at an angle of 30° to the horizontal. (See *A* in Fig. 4.)
2. Lay out the actual width, length, and height on these three lines. (See *B* in Fig. 4.)
3. Complete each surface by drawing the necessary lines parallel to the axis. (See *C* in Fig. 4.)
4. Draw in the remaining details of the object. (See *D* in Fig. 4.)

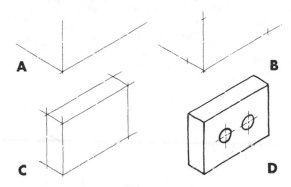

Fig. 4. These are the steps in making an isometric sketch.

5. When an object has slanted lines that do not run parallel to a main axis, these lines are all called nonisometric lines (Fig. 5). To draw nonisometric lines,

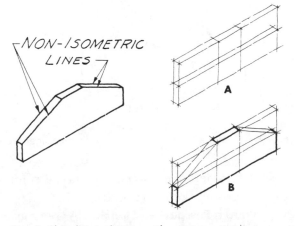

Fig. 5. Sketching objects with nonisometric lines.

first lay out one of the views as you would for a multiview sketch. Then project the points from this view to the isometric lines. The same procedure can be used to sketch angles and irregular curves.

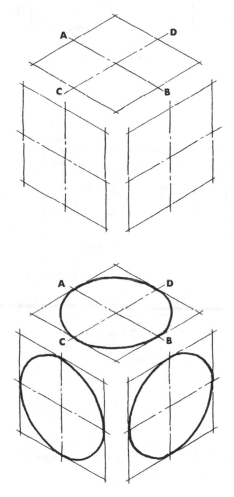

Fig. 6. This is how isometric circles are sketched.

6. To sketch an isometric circle or arc, draw an isometric square of the desired size. Draw center lines AB and CD parallel to the axes of the square. Sketch short arcs from A to C, C to B, B to D, and D to A (Fig. 6).

How To Make an Oblique Sketch. An oblique sketch is similar to an isometric sketch, except that the front face is seen in its true shape, much as the front view of a multiview sketch.

The two sides or ends are made to recede at some convenient angle from the horizontal, such as 30° or 45° (Fig. 7). The common practice is to select the face that is most irregular

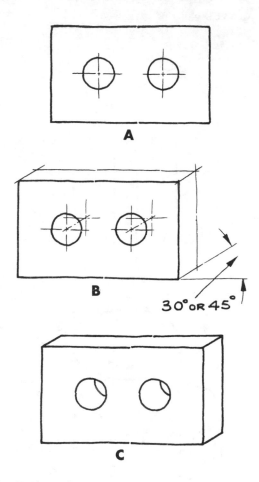

Fig. 7. This is the procedure for making an oblique sketch.

or complicated in shape to serve as the front. To avoid undue distortion, and produce a more pleasing appearance, it is advisable to place the face with long dimensions in front, instead of in a receding position (Fig. 8).

Since the receding faces of an oblique sketch are shown in their true size, the result often produces a slightly distorted appearance for some objects. This distortion can be minimized by foreshortening the receding lines. Foreshortening simply means reducing the receding lines one-half their actual size.

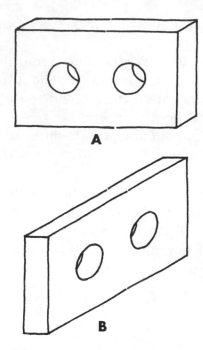

Fig. 8. Avoid distortion in oblique sketches by placing surfaces with long dimensions in front.

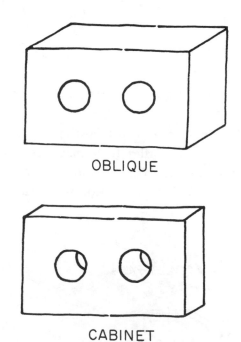

Fig. 9. An oblique sketch compared with a cabinet sketch.

A sketch, when made in such a way, is referred to as a cabinet sketch (Fig. 9).

Making a Perspective Sketch. A perspective sketch is one which more nearly presents an

Fig. 10. This is a perspective sketch.

line in the distance, and at eye level. All lines below eye level have the appearance of rising upward to the horizon, and all lines above eye level appear to go downward to the horizon. The object can be sketched so the vanishing point is either to the left or to the right, or directly in the center of vision (Fig. 11).

object as it appears to the eye or as seen in an actual picture. A perspective sketch is based on the fact that all lines which extend from the observer appear to converge or draw together at some distant point. For example, in sighting down a long stretch of highway, light poles, wires, and buildings will appear to slant and run together, as shown in Fig. 10. The point where all lines seem to meet is known as the vanishing point. This vanishing point is located on the horizon, which is an imaginary

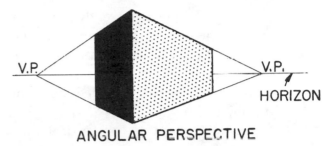

ANGULAR PERSPECTIVE

Fig. 12. A perspective sketch may have two vanishing points.

Some perspective sketches may have two vanishing points, as shown in Fig. 12. When a sketch is made with one vanishing point, it is

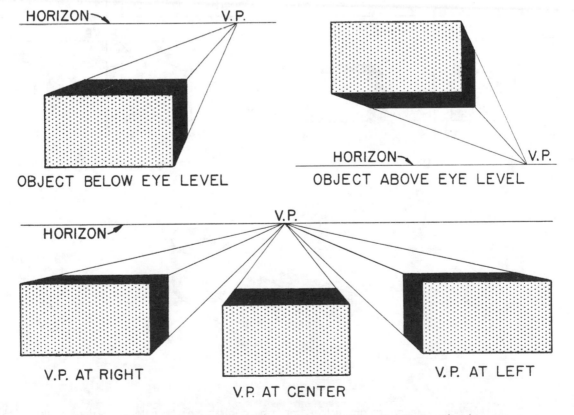

OBJECT BELOW EYE LEVEL

OBJECT ABOVE EYE LEVEL

V.P. AT RIGHT

V.P. AT CENTER

V.P. AT LEFT

Fig. 11. These are the positions of the vanishing point in perspective sketches.

said to be drawn in parallel perspective. A sketch with two vanishing points is known as an angular perspective.

To make a simple perspective sketch, proceed as follows:

1. Assume the location of the horizon line.
2. Locate the position of the vanishing points.
3. For a parallel perspective, draw a front view of the object. This will be a true view (Fig. 13). If an angular perspective is to be made, draw a vertical line and, on it, lay off the full or scaled height of the object (Fig. 14).
4. From the front view or vertical line,

draw light construction lines back toward the vanishing points.

5. The points designating the length or depth of the object may be found by a projection method. However, since this is a complicated procedure, it is used only when an accurate mechanical perspective is made. For most purposes, location points for depths are simply assumed, that is, the vertical lines representing the ends of the object are placed in a position that produces the most pleasing effect.
6. Darken *in* all outlines. Surfaces that lie in a shaded area may be lightly shaded.

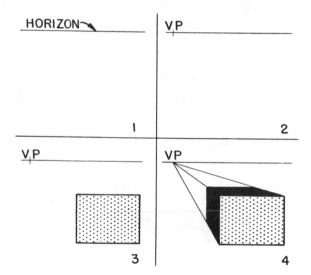

Fig. 13. Making a parallel perspective sketch.

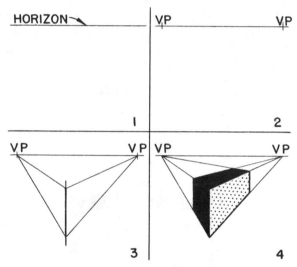

Fig. 14. Making an angular perspective sketch.

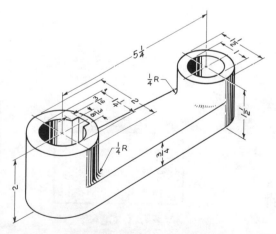

Fig. 15. When dimensions are included on a pictorial sketch, the lettering is done in the plane of measurement.

PROBLEM 1. *Construct an isometric sketch of the flanged support.*

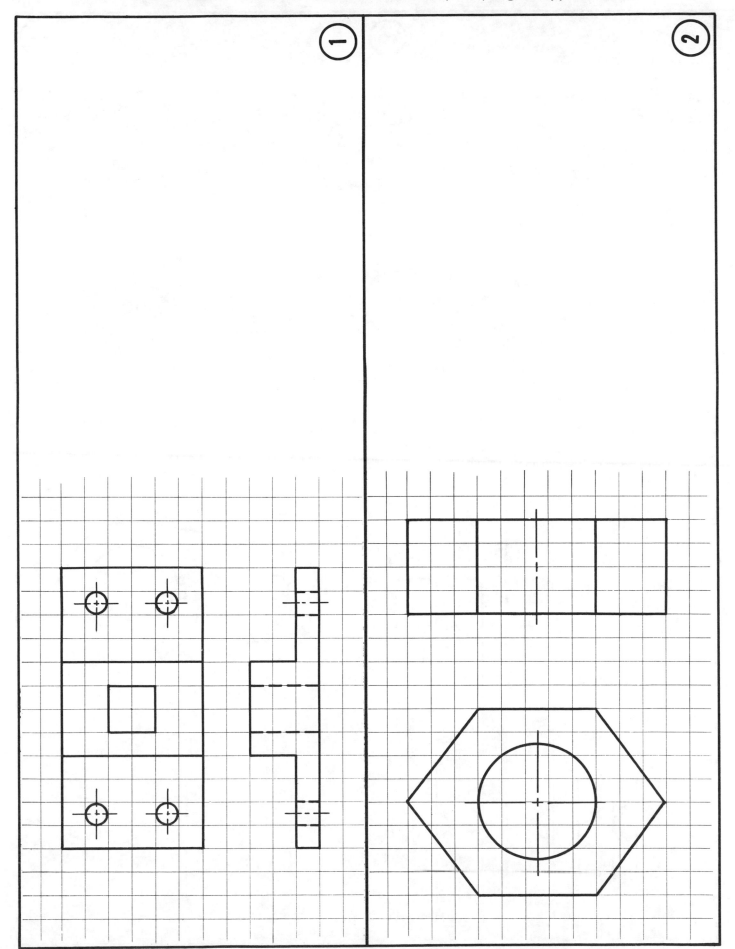

PROBLEM 2. *Construct an isometric sketch of the spacing collar.*

PROBLEM 3. *Construct an isometric sketch of the pivot. Show all dimensions.*

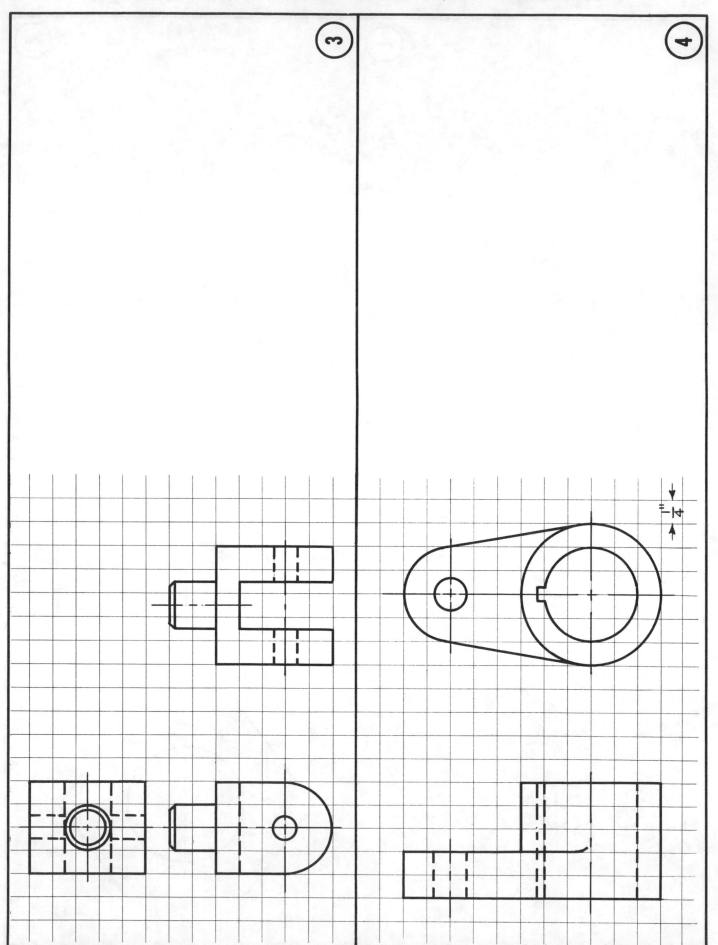

PROBLEM 4. *Construct an isometric sketch of the shear hub. Include all dimensions.*

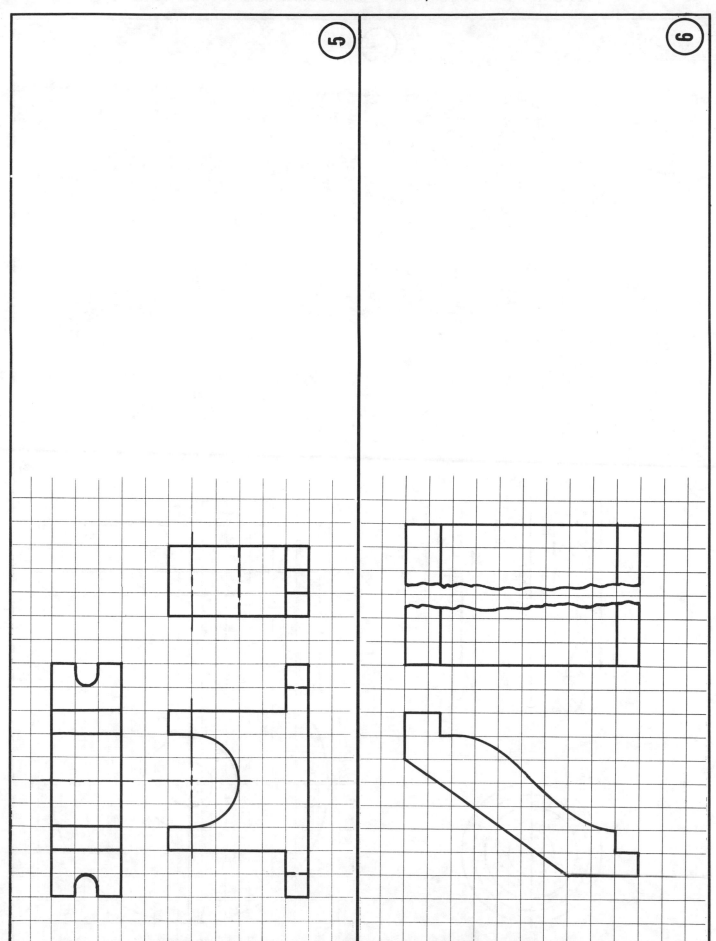

PROBLEM 7. *Prepare a full sectional cabinet sketch of the chuck.*

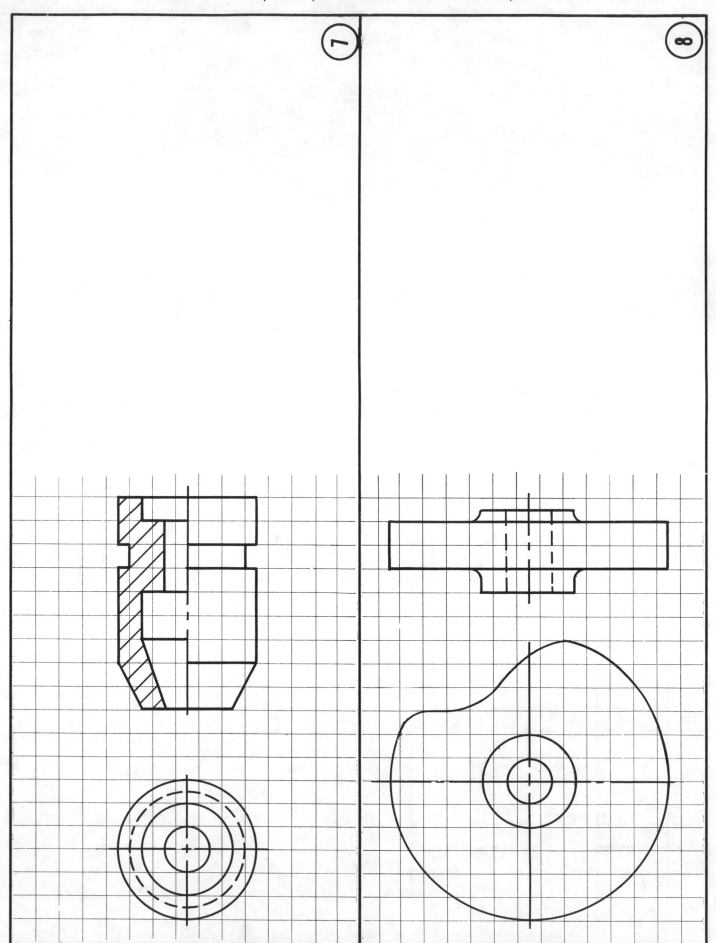

PROBLEM 8. *Make a cabinet sketch of the cam.*

PROBLEM 9. *Construct a parallel perspective of the* V*-block.*

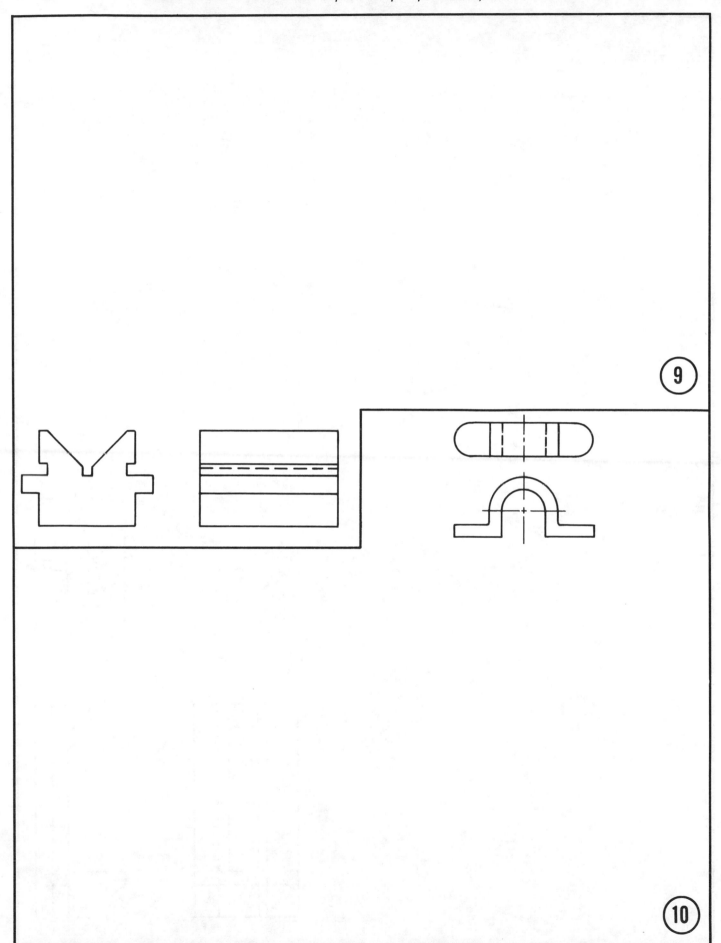

⑨

⑩

PROBLEM 10. *Prepare a parallel perspective sketch of the clamp.*

PROBLEM 11. *Make an angular perspective sketch of the vise base.*

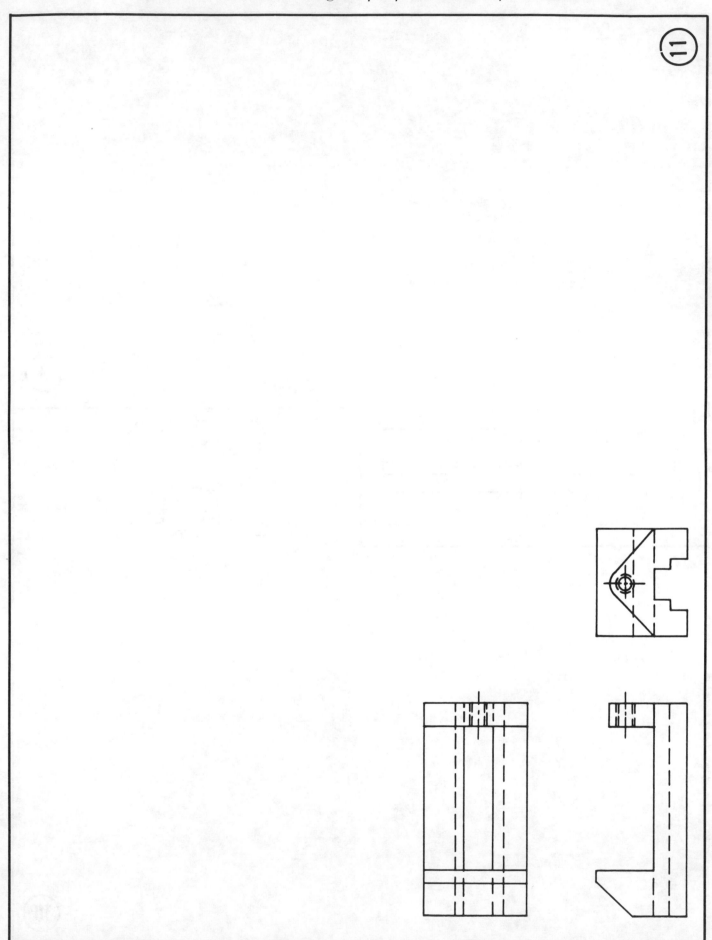

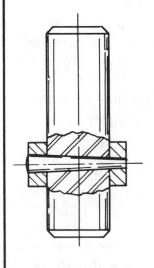

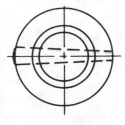

UNIT 8

SKETCHING FASTENING DEVICES

In the process of making some types of sketches, you will need to show various fasteners which are to be used in assembling parts of an object. To simplify the job of sketching these fasteners, certain conventions are used. In this unit, you will learn about some of the more common types of fasteners and how they should be drawn. See Fig. 1 for basic terms relating to screw threads.

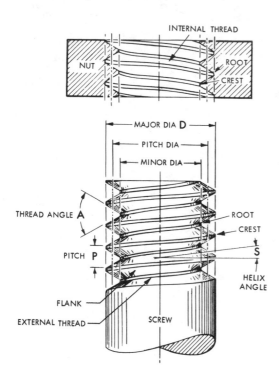

Fig. 1. Illustration of thread terms.

Unified Screw-Thread Series. The unified screw-thread series is essentially the American Standard. The difference is primarily in the degree of tolerance incorporated in the manufacturing process. This series was designed to bring about greater interchange of threaded parts between the United States and other countries. The series consists of the following types:

Unified National Coarse—UNC
Unified National Fine—UNF
Unified National Extra Fine—UNEF
Unified 8-pitch—8 UN
Unified 12-pitch—12 UN
Unified 16-pitch—16 UN

1. *Unified National Coarse threads* are used for the bulk production of screws, bolts and nuts for general applications requiring rapid assembly or disassembly.
2. *Unified National Fine threads* are used on bolts and nuts employed in the automotive and aircraft industries where greater resistance to vibration is required or where greater holding strength is necessary.
3. *Unified National Extra Fine threads* are used for threaded parts which require a fine adjustment, such as bearing retaining nuts, adjusting screws and thin walled tubing.
4. *Unified 8-pitch threads* are used in the utility industries for high-temperature bolting in steam flange connections or as substitutes for coarse threads.
5. *Unified 12-pitch threads* are used in machine construction for thin nuts on shafts and sleeves. They are also used as a continuation of the fine series for diameters larger than 1½ inch.
6. *Unified 16-pitch threads* are used for adjusting collars and retaining nuts. They also serve as a continuation of the extra-fine series for diameters larger than 1-11/16 inch.

Thread Classes. Thread class, which formerly was referred to as fit of a thread, describes the degree of looseness or tightness between two mating threaded parts. There are three classes in the United Thread Series. These classes are designated by numbers followed by the letter *A* for external threads and *B* for internal threads.

1. *Classes 1A and 1B* have the greatest amount of allowance and are intended where application requires frequent and rapid assembly and disassembly with minimum binding.

2. *Classes 2A and 2B* are considered standard for general purpose threads on bolts, nuts and screws. They are widely used for mass production work.

3. *Classes 3A and 3B* are suitable where very close tolerances are required such as aircraft bolts, set screws and cap screws.

How Threads Are Drawn. Threads on any fastener are drawn by using either the *schematic* or *simplified* symbol (Fig. 2). To produce a regular thread symbol, crest lines representing the number of threads are uniformly spaced by eye. These lines should not be closer than $\frac{1}{16}$″; the actual number of threads need not be shown. The root lines are centered between the crest lines and terminated a short distance from the outside diameter of the thread. The simplified symbol is made by merely drawing two invisible lines representing the root of the thread parallel to the axis.

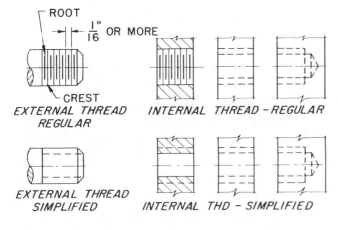

Fig. 2. This is how threads are drawn.

Designating Screw Threads. On a drawing a screw thread is designated by a note with a leader and arrow pointing to the thread (Fig. 3).

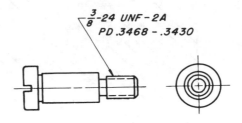

Fig. 3. The way a thread is designated on a drawing.

Types of Screw-Threaded Fasteners. The following paragraphs list the most common types of screw-threaded fasteners.

Bolts. The term *bolt* applies to a fastener having a head on one end and a stem that fits through a drilled opening. The opposite end of the bolt is threaded to receive a nut which can be tightened to hold parts together (Fig. 4). Bolts and matching nuts have either square or hexagonal heads.

Studs. A stud, or stud bolt, is a rod threaded

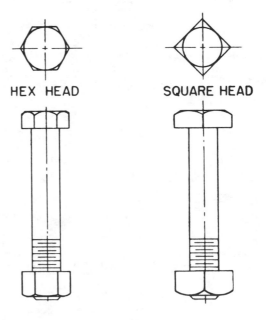

Fig. 4. These are Standard machine bolts.

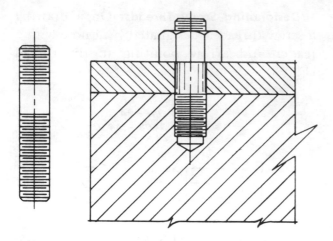

Fig. 5. This shows how a stud is used.

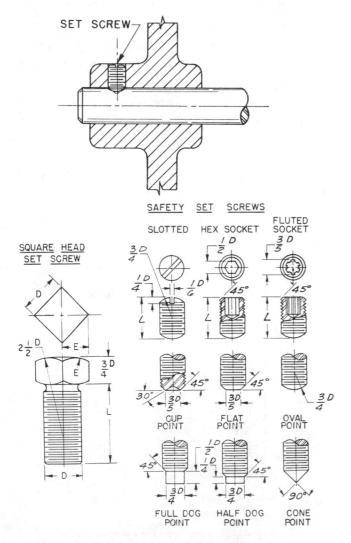

Fig. 6. These are the types of set screws commonly used.

on both ends (Fig. 5). It is used when regular bolts are not suitable, especially on parts that must be removed frequently, such as cylinder heads. One end of the stud is screwed into a threaded or tapped hole, and the other end usually fits into the removable piece of the structure. A nut is used on the projecting end to hold the parts together.

Set Screws. The function of a set screw is to prevent rotary motion between two parts, such as the hub of a pulley and a shaft (Fig. 6). The set screw is driven into one part so that its point bears firmly against another part. Set screws are either headless or have a square head. They are available with a variety of points, as shown in Fig. 6.

Cap Screws. A cap screw passes through a clearance hole in one member of the structure and screws into a threaded or tapped hole in

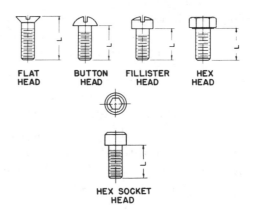

Fig. 7. These are cap screws.

the other. They range in diameter from ¼" to 1¼" and are available in five standard head types, as shown in Fig. 7.

Machine Screws. Machine screws are similar to cap screws except that they are smaller and are used chiefly on small work having thin sections. Below ¼" size machine screws are specified by numbers for 2 to 12. Above ¼" the size is indicated by diameter.

Stove Bolts. Stove bolts are used for joining parts whenever great strength is not impor-

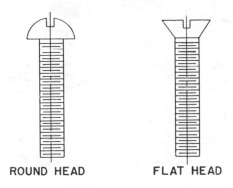

ROUND HEAD   FLAT HEAD

Fig. 8. These are stove bolts.

three types of heads: flat, oval, and round (Fig. 10.). Fig. 11 shows how screws are sketched.

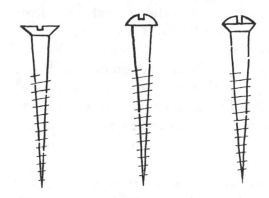

Fig. 11. This is how wood screws are sketched.

tant (Fig. 8). These bolts are available with a round or flat head. All heads are slotted so a screw driver can be used to tighten them.

Carriage Bolts. Carriage bolts are used to fasten two pieces of wood together, or one section of wood to a metal member (Fig. 9). This bolt has a square section directly under an oval head. The square part prevents the

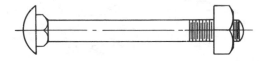

Fig. 9. This is a carriage bolt.

bolt from turning when drawn into the wood while the nut is being tightened.

Wood Screws. Wood screws are made of steel, brass, and bronze. They are available in

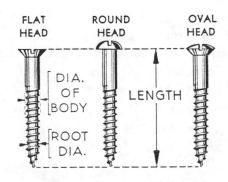

FLAT   ROUND   OVAL
HEAD   HEAD   HEAD

DIA.
OF
BODY   LENGTH

ROOT
DIA.

Fig. 10. Here are the common types of wood screws.

Rivets. Rivets are considered as permanent fastening devices and are used in joining parts constructed of sheet metal or plate steel. They are made of many different kinds of metal. The most common are wrought iron, soft steel, copper, brass, and aluminum. Rivets are available with a flat, countersunk, button (often referred to as round or oval head), pan, and truss type of head (Fig. 12).

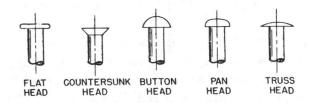

FLAT   COUNTERSUNK BUTTON   PAN   TRUSS
HEAD   HEAD   HEAD   HEAD   HEAD

Fig. 12. These are common types of rivets.

Nails. There are many different kinds of nails, such as common, box, finishing, casing and brad. *Common* nails have larger diameters and wider heads than other types. They are used mostly in rough carpentry. *Box* nails also have wide heads but are not as large in diameter as common nails. They are used extensively in box construction and in many types of carpentry where common nails would be unsuitable. *Casing* nails are smaller in diameter and head size than box nails. They are used especially in blind nailing of flooring and ceiling, and cabinet work where large heads are undesirable. *Finishing* nails have the smallest diameter and the smallest heads. They are used in cabinet work and furniture construction where it is often necessary to sink the heads below the surface of the wood.

Sizes of nails are designated by the term penny (d), with a number as a prefix such as 4d, 10d. The term penny refers to the weight of the nail per thousand in quantity. A 6d nail means that the nails weigh six pounds per thousand.

Brads are the smallest type of nails. The sizes of brads are indicated by the length in inches and the diameter by the gauge number of the wire. The higher the gauge number, the smaller the diameter.

Washers and Locking Devices. The two main types of washers are plain and lock washers. *Plain* washers are used primarily as a bearing surface. *Lock* washers serve as a locking device to prevent nuts from becoming loose under vibrations (Fig. 13).

In addition to lock washers, other devices are often used to secure nuts. The *castle* and *slotted nuts* are designed so a cotter key can be inserted in the slotted nut-and-bolt assembly. The *Palnut* has a slotted, cone-shaped, thread-engaging section which, when tightened, is forced against the regular nut. The *elastic stop nut* has a red fiber-locking collar that is slightly smaller than the diameter of the bolt. As the nut is screwed on, the fiber forms to the bolt thread and grips it tightly, to prevent the nut from coming loose. The *jam nut* is simply another nut, thinner in section than a regular nut, that is screwed up tight against the regular nut.

How Threaded Fasteners Are Specified: On a drawing it is necessary to show the specifications of the fastener used. This information is listed as follows:

Bolts and Studs. ⅜—16NC—2 Hex Hd. Bolt × 2½ Lg. or ⅜—16UNC—2A Hex Hd. Bolt × 2½ Lg.

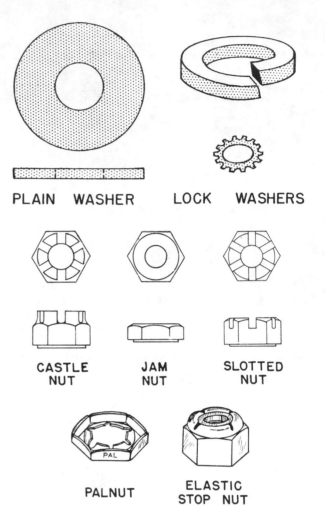

PLAIN WASHER   LOCK WASHERS

CASTLE NUT   JAM NUT   SLOTTED NUT

PALNUT   ELASTIC STOP NUT

Fig. 13. These are the types of washers and nuts most generally used.

Set Screws. ¼—20NC—2 Sq. Hd. Oval Pt. Set Screw × ½ Lg. *or* ¼—20UNC—2A Sq. Hd. Oval Pt. Set Screw × ½ Lg.

Cap Screws. ½—13NC—3 Button Hd. Cap Screw × 1 Lg. *or* ½—13 UNC—3A Button Hd. Cap Screw × 1 Lg.

Machine Screws. #10—24NC—2 Rd. Hd. Mach. Screw × ⅜ Lg. *or* #10—24UNC—2A Rd. Hd. Mach. Screw × ⅜ Lg.

Stove Bolts. ³⁄₁₆ Rd. Hd. Stove Bolt × 1 Lg.

Carriage Bolts. ¾—Carriage Bolt × 4 Lg.

Wood Screws. 1½ #8, F.H. Steel Wood Screw.

Nails. #10D Common Nail.

Rivets. ⅛DIA × 1 Rd. Hd. Steel Rivet.

Welding. Welding is a very common method of joining parts and therefore should be considered as a fastening device. A system of designating welded joints or drawings has been standardized by the American Welding Society and adopted by most industries.

Fig. 14 illustrates basic joints for assembling metal units. Fig. 15 shows the different kinds of common welds that are used to join

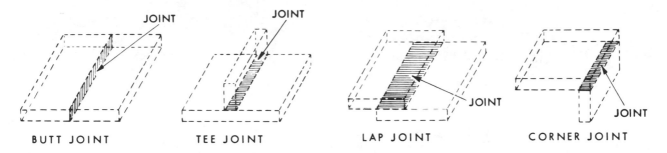

Fig. 14. Basic weld joints. American Welding Society.

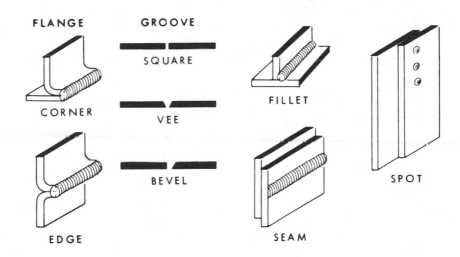

Fig. 15. Common welds used in joining metal parts. American Welding Society.

FLASH OR UPSET	GROOVE							FILLET	PLUG OR SLOT	SPOT OR PROJECTION	SEAM
SQUARE	V	BEVEL	U	J	FLARE-V	FLARE-BEVEL					
‖	∨	⋁	⊌	⊍	)(	⎰⎱	◺	▭	○	⊖	

BACK OR BACKING	SURFACING	FLANGE		WELD ALL AROUND	FIELD WELD	MELT THRU	CONTOUR		
		EDGE	CORNER				FLUSH	CONVEX	CONCAVE
⌣	⌣⌣	⏝	⎿	⊙	●	⬤	—	⌒	⌣

Fig. 16. Symbols used to designate welds. American Welding Society.

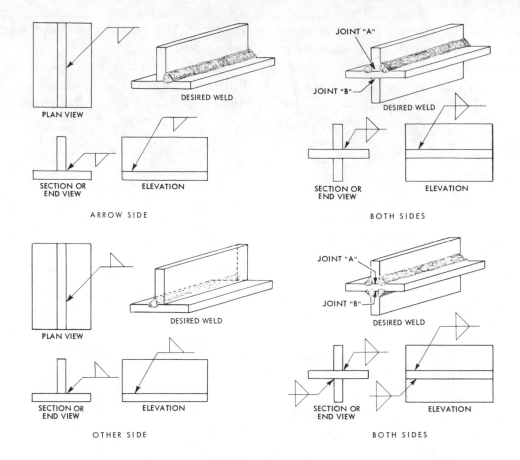

Fig. 17. Location of weld symbols. American Welding Society.

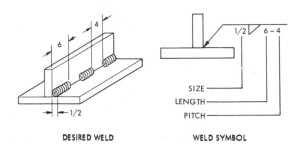

Fig. 18. How sizes of welds are indicated.

metal parts, and Fig. 16, the symbols used to designate welds.

Fig. 17 presents how weld symbols are placed on a drawing. Along with weld symbols, figures are also used to indicate the sizes of welds required. Some of the more common weld specifications are shown in Fig. 18.

PROBLEM 1. *On center lines* A, *fasten the two parts of the clamp assembly, using Round Head Machine Screws with square nuts. Give complete specifications for the fasteners.*

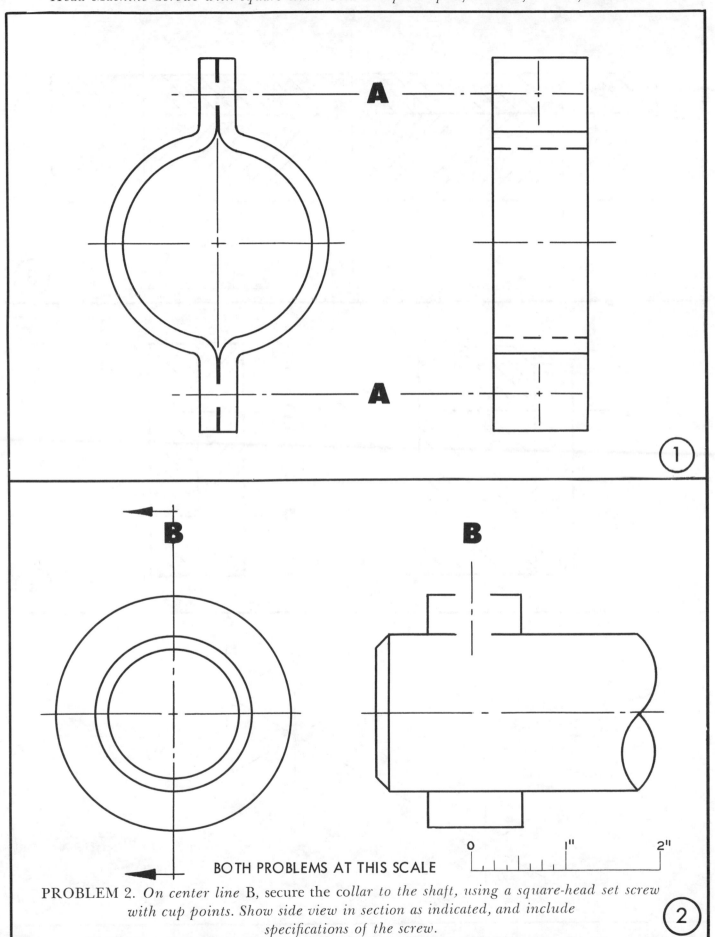

BOTH PROBLEMS AT THIS SCALE

0   1"   2"

PROBLEM 2. *On center line* B, *secure the collar to the shaft, using a square-head set screw with cup points. Show side view in section as indicated, and include specifications of the screw.*

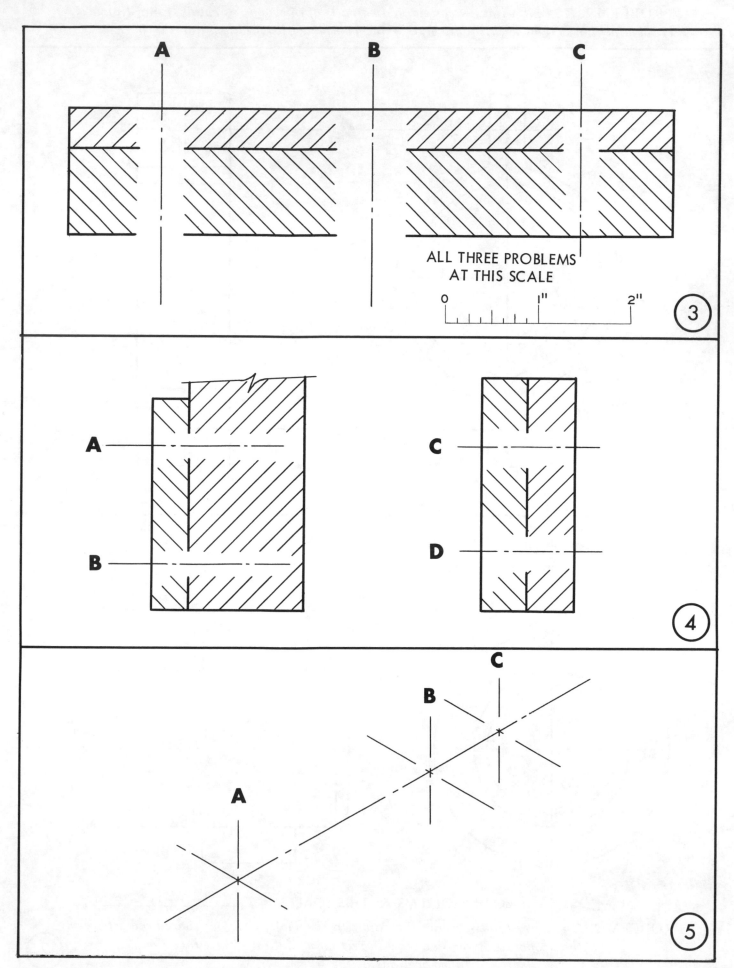

ALL THREE PROBLEMS
AT THIS SCALE

0   1"   2"

③

④

⑤

PROBLEM 3. *Fasten the two pieces of stock with the following:*

At A, use ½–13UNC–2A Hex. Hd. Bolt × 2 Lg. with Washer and Hex. Nut
At B, use ¾–10UNC–2A Sq. Hd. Bolt × 2½ Lg. with Washer and Sq. Nut
At C, use ⅜–16UNC Stud × 1¼ Lg. with Hex. Nut

PROBLEM 4. *Fasten parts using the following:*

At A, use ¼–20UNC Flat Hd. Mach. Screw × 1 Lg.
At B, use ¼–20UNC Rd. Hd. Mach. Screw × 1¼ Lg.
At C, use ⅜ D × 1¼ Button Hd. Rivet.
At D, use ⁵⁄₁₆ D × 1¼ Csk. Rivet.

PROBLEM 5. *Make an isometric sketch, showing a standard Hex. head bolt at A, a stand-ard plain washer at B, and a standard Hex. nut at C.*

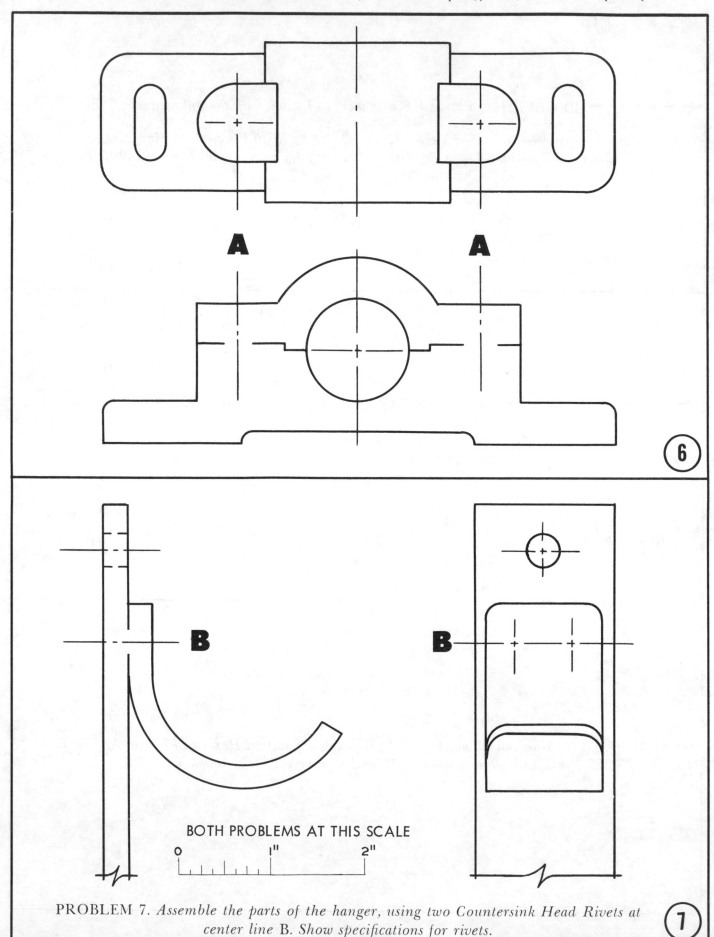

PROBLEM 6. *Complete the assembly of the Pillow Block Bearing, using Hex. Head Cap Screws at center lines A. Show front view in full section. Specify the fasteners completely.*

A   **A**

⑥

B

B

BOTH PROBLEMS AT THIS SCALE

0   1"   2"

PROBLEM 7. *Assemble the parts of the hanger, using two Countersink Head Rivets at center line B. Show specifications for rivets.*

⑦

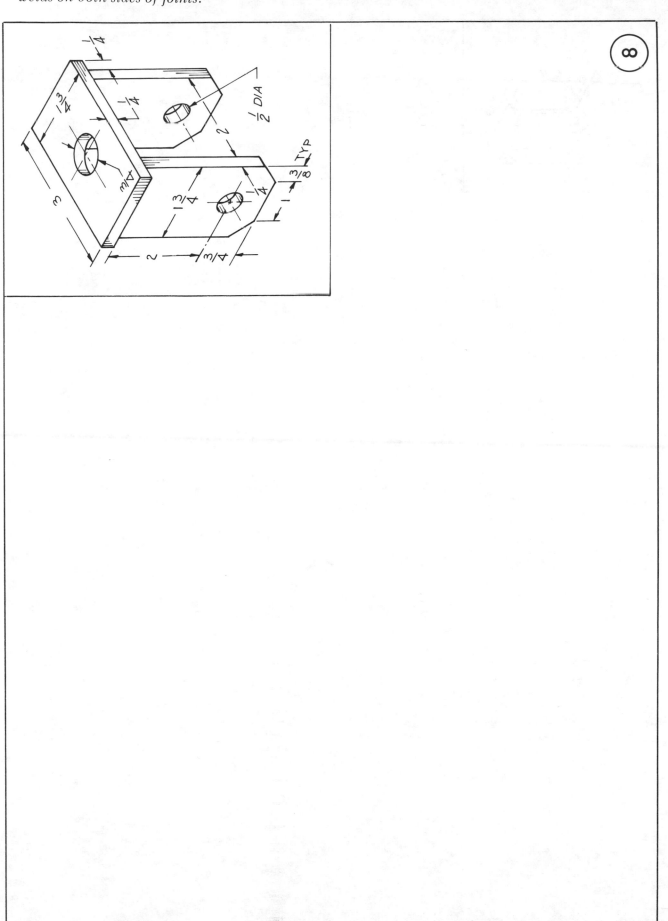

8

PROBLEM 9. *Make a three-view sketch and dimension. Show pieces welded with appropriate-sized fillet weld on one side.*

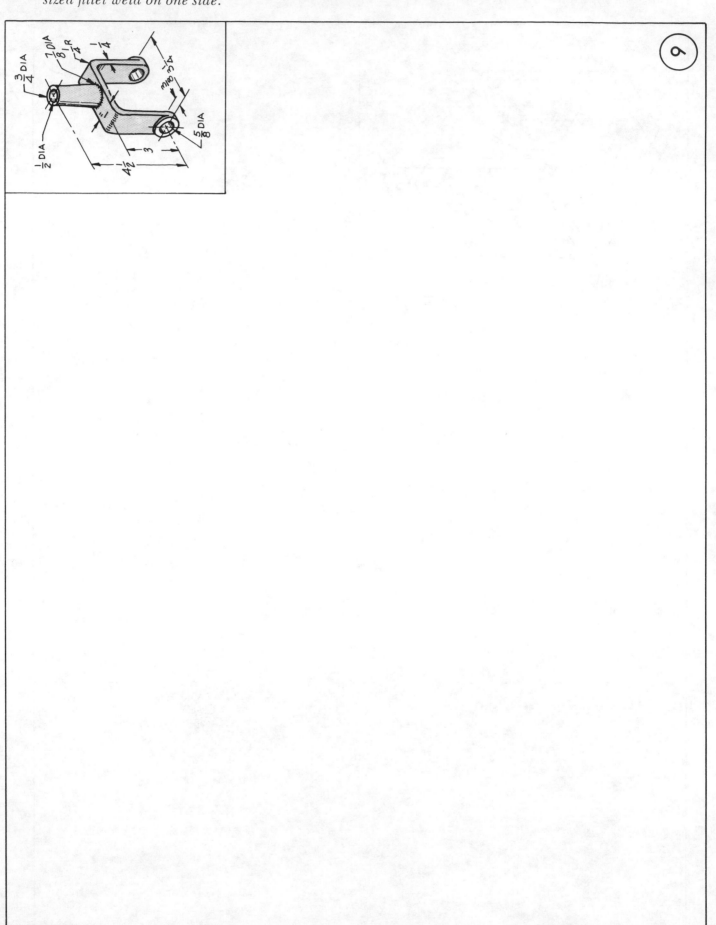

UNIT 9

SHADING A SKETCH

One of the main purposes of a pictorial sketch is to achieve better visualization of an object. It is a means of bridging the gap between an actual photograph and the natural object. To make a pictorial sketch appear more natural, a process of shading is often used. Shading is simply a technique of varying the light intensity on the surfaces by lines or tones. Notice in Fig. 1 how shading helps to better visualize the shape of the object.

Location of Shaded Areas. Since shading is a result of light intensity on a surface, the first consideration in producing a shaded effect is to determine the source of light falling on the object. Generally speaking, one can proceed on the basis that the principal source of light is shining over the observer's left shoulder or from the upper corner of the drafting board, as shown in Fig. 2. This, however, is not necessarily a fixed rule since sources of light from

Fig. 1. Shading aids the imagination in visualizing the shape of the object.

Fig. 2. Direction of the main source of light falling on an object.

95

other directions may often produce an even more pleasing appearance. However, if we assume that the standard direction of light is over the left shoulder, then the top and front surfaces of the object receive the most light. The sides that form the smallest angle with the light rays have less light, and the surfaces that are directly opposite the light source are in deep shade. (Fig. 3).

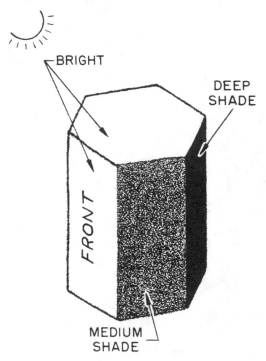

Fig. 3. How the standard direction of light affects the surfaces of an object.

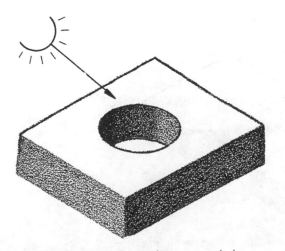

Fig. 4. Place a shaded area adjacent to a light area.

By following this rule, shading can be accomplished by sketching contrasting weights of lines over the surfaces affected by the light. It is well to keep in mind, too, that a more pleasing effect can be obtained by having the shaded area of one plane adjacent to the light area of an adjoining surface (Fig. 4).

Before actually proceeding with any shading, it is good practice to outline with a very fine line the areas that are to be shaded. The line should be light enough so it can be erased after the shading is completed (Fig. 5).

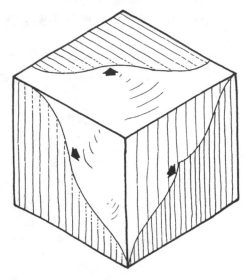

Fig. 5. Sketch the outline of the shaded areas with a fine line.

Line Shading. The simplest method of producing shading effects is by means of contrasting weights and spacing of lines. Fig. 6 shows how such lines are sketched on flat surfaces. The spacing of lines should be judged by eye. Notice that the darker the area, the closer will be the lines. Actually no hard and fast rule can be given as to the amount of space to leave between lines. Practice and judgment will serve as the best guide. The point to keep in mind is to visualize the intensity of light cast on the object and then space the lines so they will best reflect the effects of this light.

The weight of lines can be achieved by varying the pressure on the pencil. Notice in

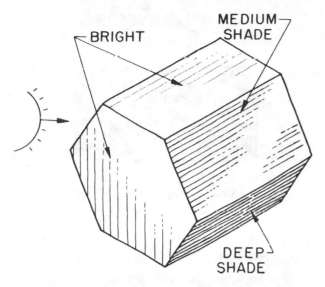

Fig. 6. This shows line shading on flat surfaces.

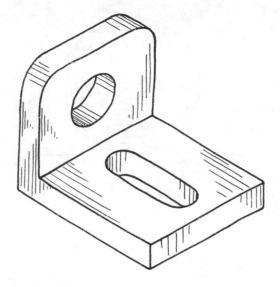

Fig. 8. This shows the direction of lines for the best shading effects.

Fig. 6 that the heaviest lines are used where the surface is to have the darkest areas. It is also possible to produce shaded effects by keeping all the lines light and varying the spacing (Fig. 7). As a rule, better results will be achieved by varying both the spacing and weight of lines.

The direction of the lines must also be taken into consideration. Usually, it is best to shade vertical faces with vertical lines and the other faces with lines parallel to one of the edges of the object (Fig. 8).

To shade curved surfaces, the lines may be sketched straight or curved. The practice is to shade approximately one-fifth of the surface nearest the source of light, then leave the next two-fifths white and shade the two-fifths that are the farthest from the light (Fig. 9). Spheres

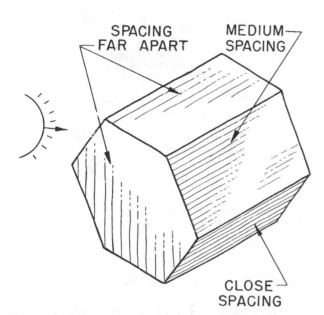

Fig. 7. Shading can be produced by varying the spacing of lines.

Fig. 9. This is how a curved surface can be shaded.

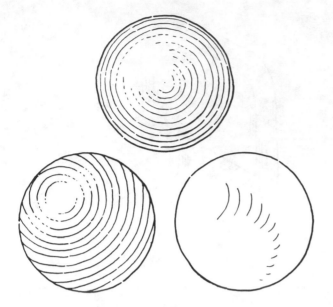

Fig. 10. This is how to shade a sphere.

Fig. 12. This is an example of broad stroke shading.

are shaded by sketching a series of concentric circles, as illustrated in Fig. 10.

Stippling. Stippling is another method which can be used to produce shaded areas.

the desired areas with the flattened side of the lead pencil (Fig. 12).

If a smudge effect is desired, as shown in Fig. 13, the broad pencil strokes should be rubbed with a paper stump. Once the area is

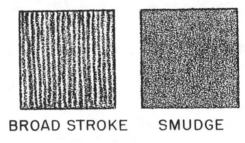

BROAD STROKE   SMUDGE

Fig. 13. This is an example of smudge shading.

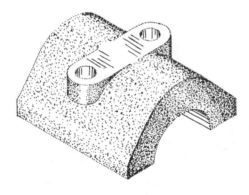

Fig. 11. Shaded areas can be achieved by stippling.

This method consists of covering the surface to be shaded with a series of dots made with the point of the pencil. For dark areas the dots are placed closer together, and for light areas the dots are spaced widely apart (Fig. 11). Although stippling produces a pleasing appearance, the process is much slower than line shading.

Broad Stroke and Smudge Shading. Good shading results can be achieved by rubbing

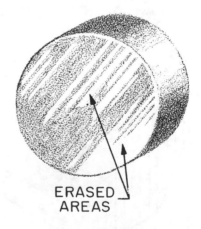

ERASED AREAS

Fig. 14. Erasing in a shaded area produces light spots.

covered, light and dark effects can be achieved by placing an erasing shield over the shaded part and rubbing with an eraser (Fig. 14).

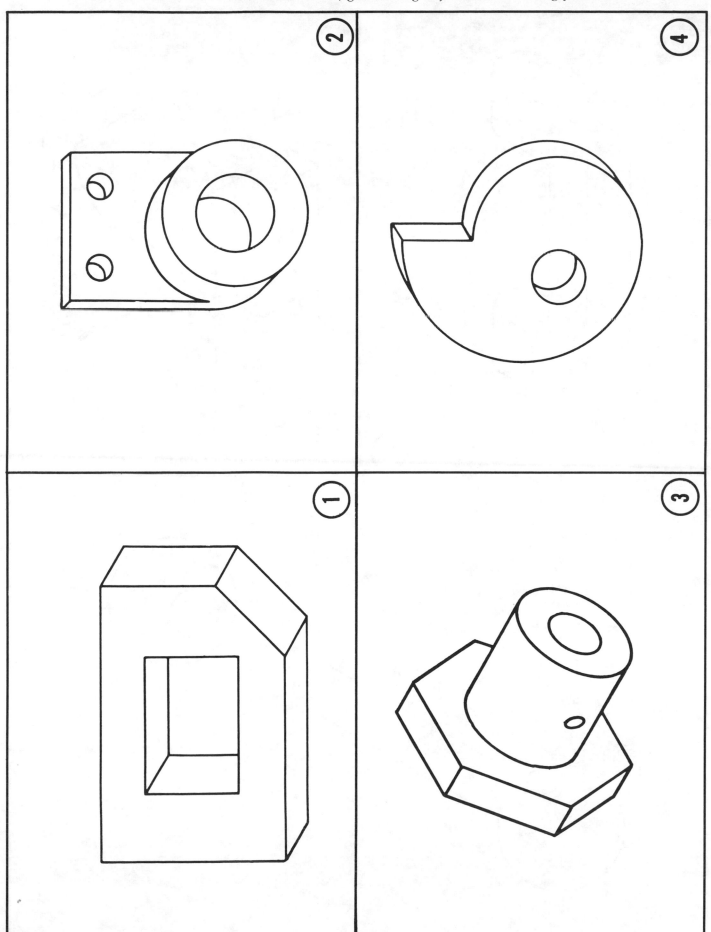

PROBLEMS 5, 6, and 7. *Shade each figure, using any suitable shading procedure.*

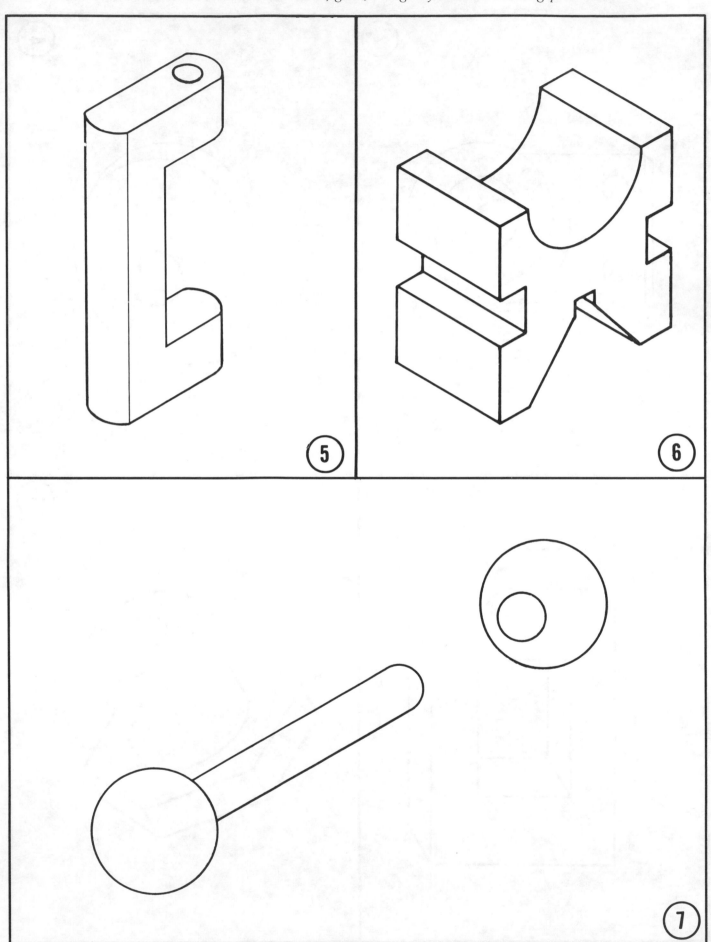

⑤

⑥

⑦

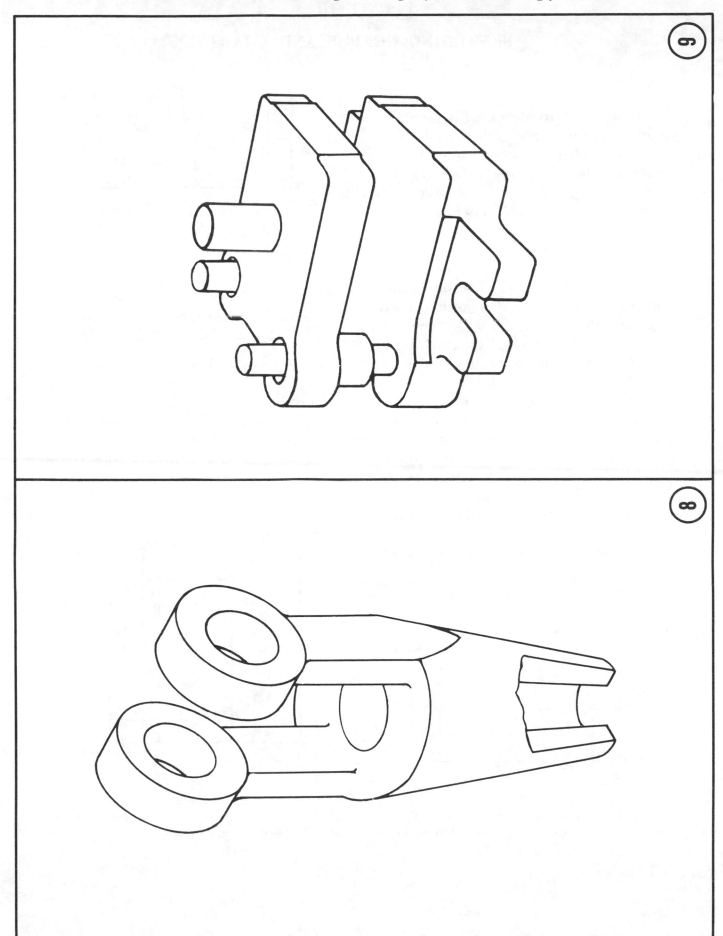

UNIT 10

SKETCHING GRAPHS AND CHARTS

Graphs and charts are used extensively in industry, business and other commercial enterprises. They are designed to present statistical data in a more interesting and meaningful form. Some of the basic types of graphs and charts are presented in this unit.

Types of Graphs and Charts. In general, a graph is said to be a pictorial presentation of numerical data for the purpose of showing comparisons, trends, or relationships. There are many different types of graphs. The most common are line graphs, bar graphs, and area graphs. To some extent, all of the others are variations of these three types.

A chart is a pictorial representation of factual material in outline form. Its purpose is to show such things as the ingredients of a product, the flow of material in a manufacturing process or the personnel organization of an office, plant or school. Actually, any form that graphically illustrates subject matter can be considered a chart.

Line Graph. A line graph consists of a single

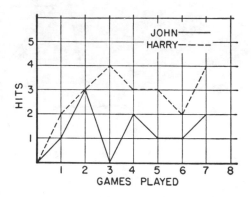

Fig. 2. The line graph may be used for comparisons. Note the effective use of different kinds of lines.

line or a series of continuous lines plotted over a specific area. Fig. 1 shows a typical line graph. Fig. 2 illustrates a type of line graph to show comparisons between two similar quantities. Very often the practice is to use different color lines for each set of figures.

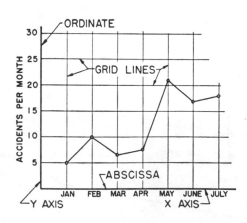

Fig. 1. A typical line graph is shown here. The terms applied to the various parts of the line graph are shown also.

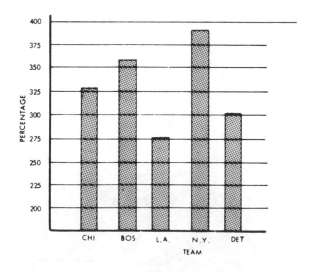

Fig. 3. A typical bar graph showing the standing of baseball teams. Different kinds of graphs should be considered before choosing the one which will best do the job.

102

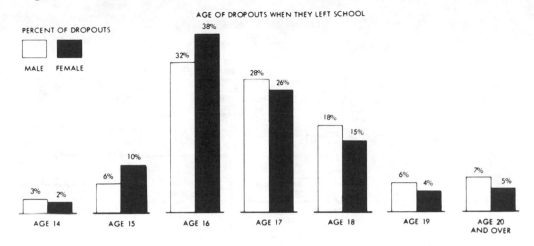

Fig. 4. This bar graph uses double bars to illustrate the comparative male-female percentages of school dropouts.

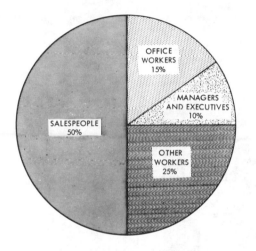

Fig. 5. The pie area graph is one of the most common of the graphs used to show a percentage breakdown of some particular quantity.

Bar Graph. A bar graph consists of a series of horizontal or vertical bars. The length of each bar represents the value or amount of a quantity being illustrated. Figures 3 and 4 are common types of bar graphs.

Area Graph. An area graph is used to show amounts or percentage breakdown of some particular quantity. Although the area graph may assume many different shapes, the most common one is the circle or pie as shown in Fig. 5. In this type the circle is simply divided into a number of segments with each portion representing a value.

Charts. Charts are used to illustrate factual material in a pictorial form. They may take any one of many different forms. Very often,

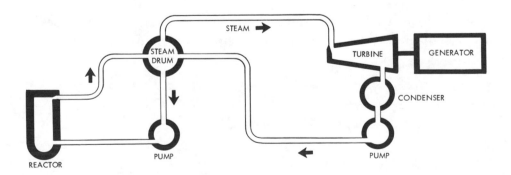

Fig. 6. A flow chart conveys a pictorial description of movements, travel or flow of some particular process or operation. Socony Mobil Oil Co.

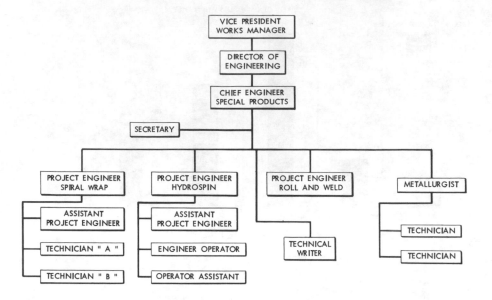

Fig. 7. An organization chart describes the relationship of a group of people or offices in an organization

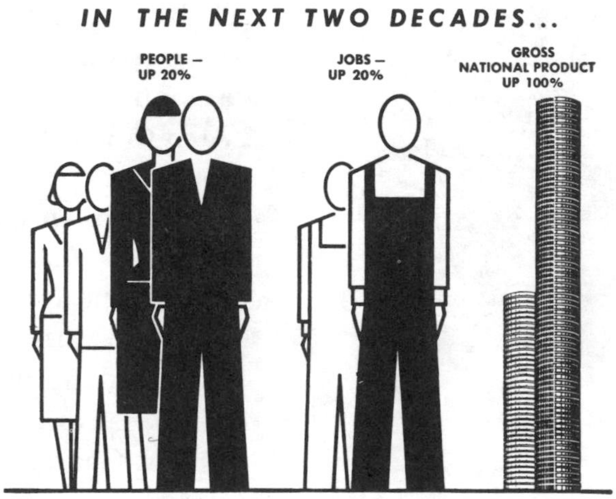

IN THE NEXT TWO DECADES...

Fig. 8. A pictorial chart uses realistic figures to present factual data.

eye-catching symbols as well as the use of color serve to present the material in a highly vivid manner. Three types of charts are shown in Figures 6, 7, and 8.

TEMPERATURE	HOUR
60°	7 A.M.
65°	8 A.M.
68°	9 A.M.
70°	10 A.M.
74°	11 A.M.
80°	12 M.
84°	1 P.M.
86°	2 P.M.
90°	3 P.M.
70°	4 P.M.
68°	5 P.M.
60°	6 P.M.
55°	7 P.M.

①

SIZE OF COMMON NAILS	LENGTH OF NAIL
6d	2"
8d	2 1/2"
9d	2 3/4"
10d	3"
12d	3 1/4"
16d	3 1/2"
20d	4"
30d	4 1/2"
40d	5"

②

PROBLEM 3. *Draw a pie graph to show the assets of the following manufacturing groups. Convert values into percentages.*

	$ MILLION
Petroleum	3590
Rubber	490
Steel	1890
Automotive	1490
Aerospace	1290
Chemicals	1090
Electrical	790

③

PROBLEM 4. *Show an organizational chart of your school, plant, office or business.*

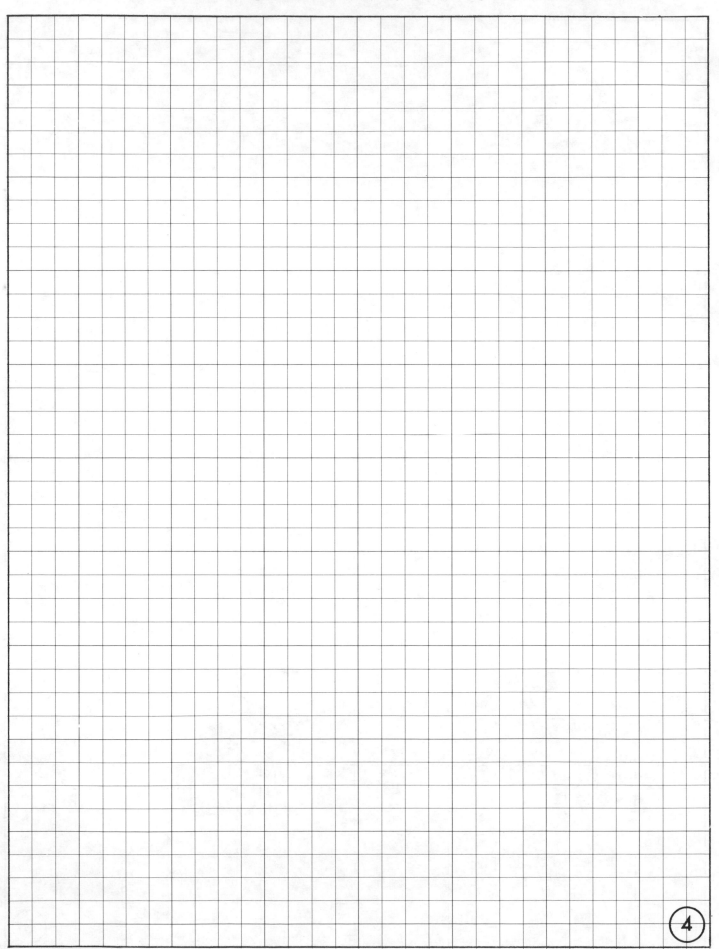

4

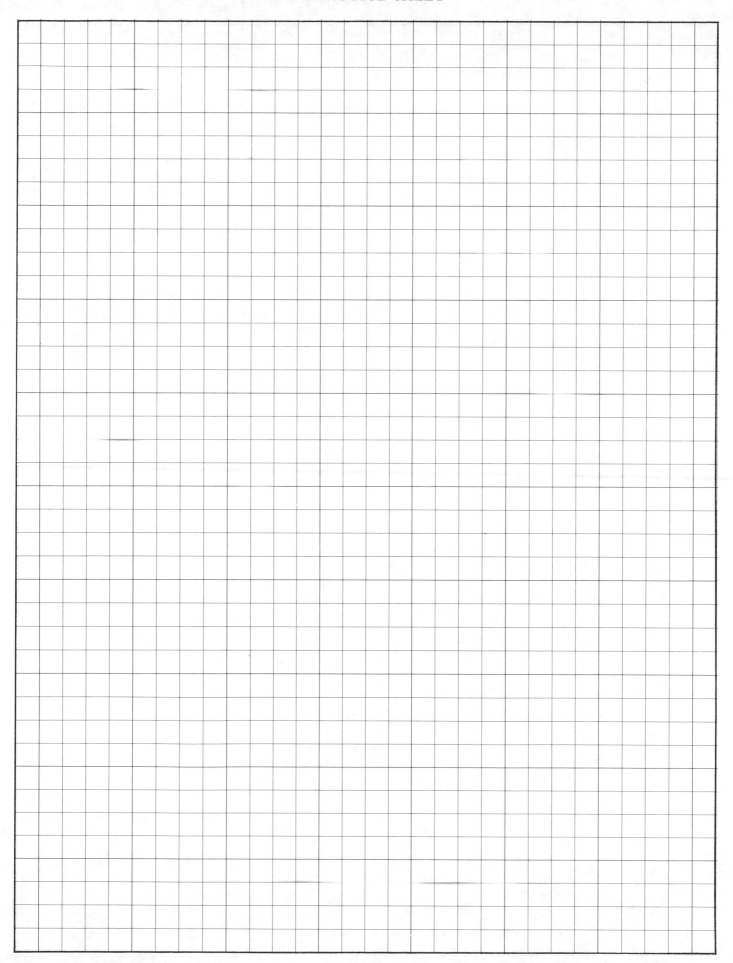

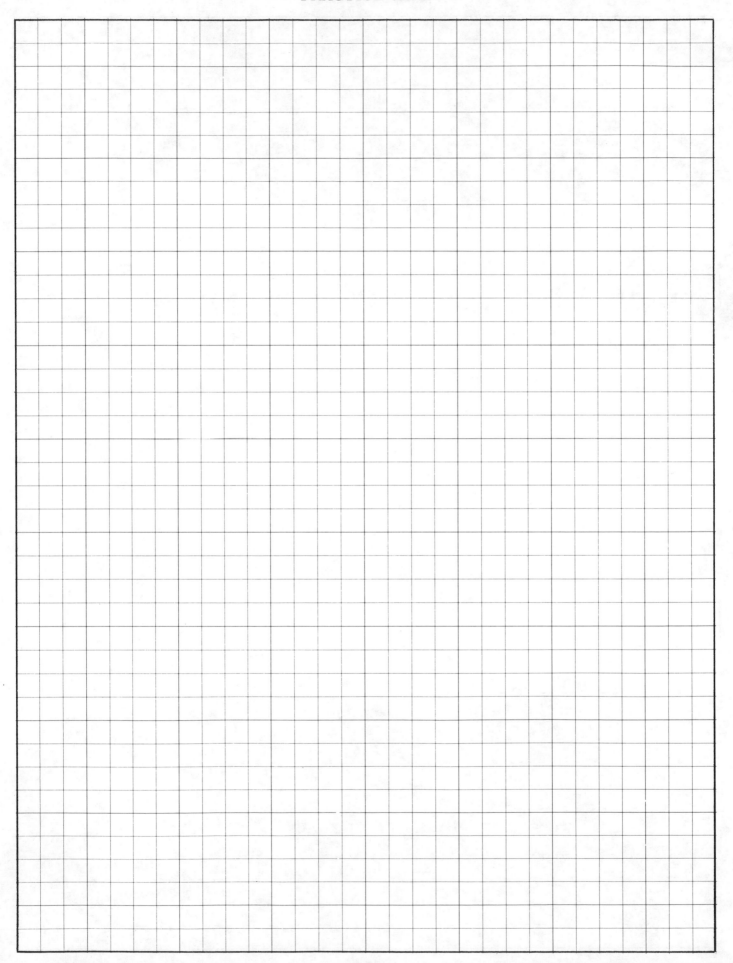

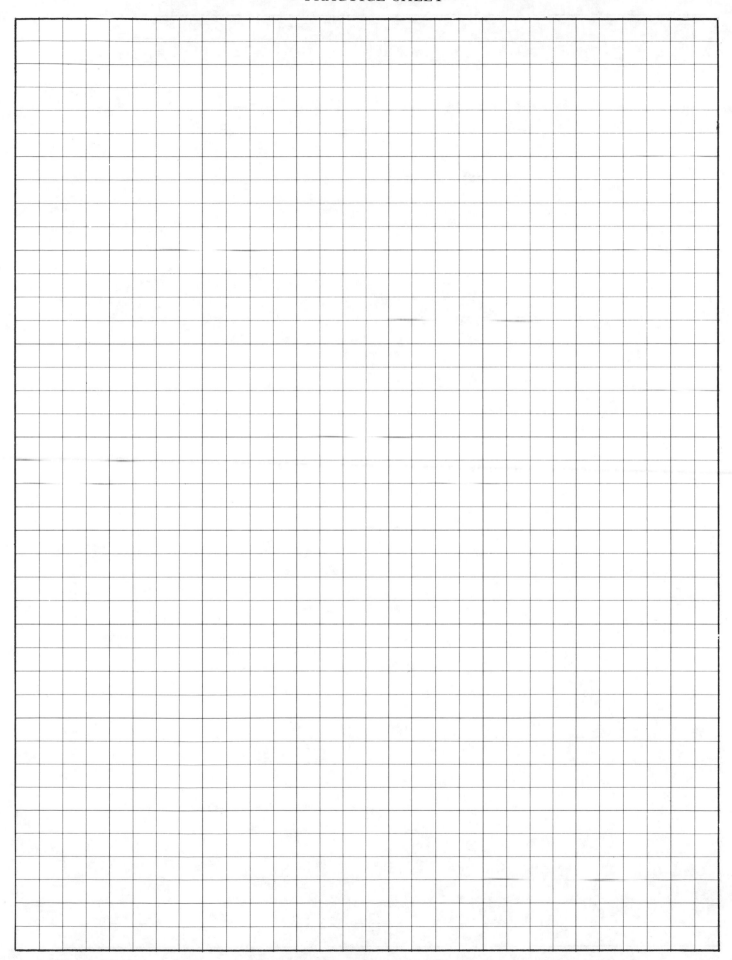

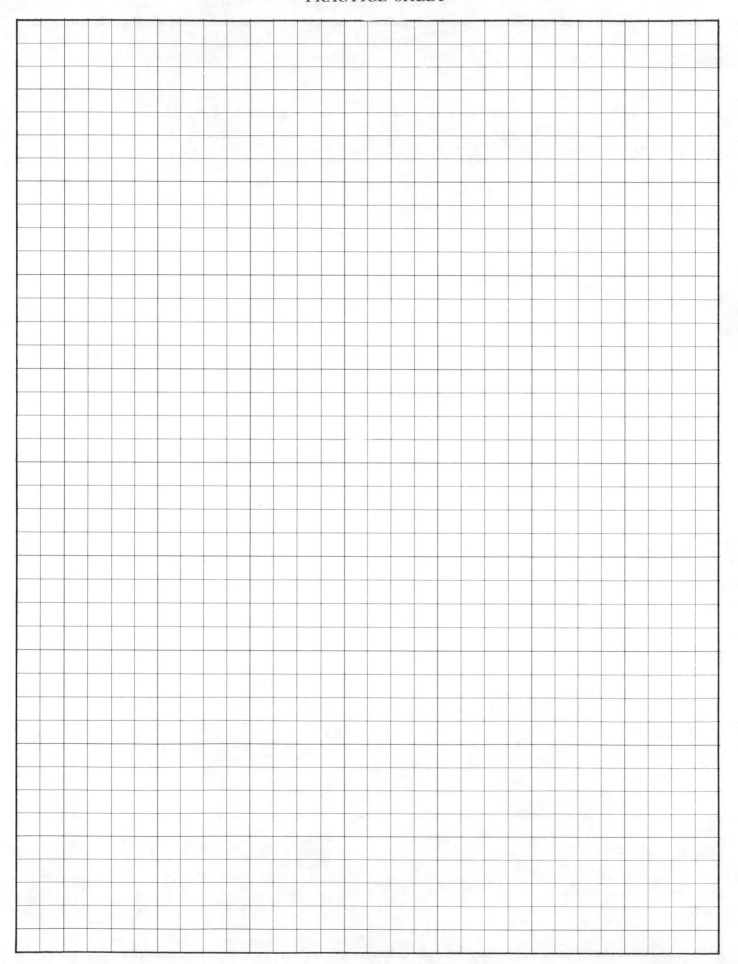

FRACTION TO DECIMAL CONVERSION CHART*

4ths	8ths	16ths	32nds	64ths	to 2 places	to 3 places	to 4 places	4ths	8ths	16ths	32nds	64ths	to 2 places	to 3 places	to 4 places
				1/64	0.02	0.016	0.0156					33/64	0.52	0.516	0.5156
			1/32		0.03	0.031	0.0312				17/32		0.53	0.531	0.5312
				3/64	0.05	0.047	0.0469					35/64	0.55	0.547	0.5469
		1/16			0.06	0.062	0.0625			9/16			0.56	0.562	0.5625
				5/64	0.08	0.078	0.0781					37/64	0.58	0.578	0.5781
			3/32		0.09	0.094	0.0938				19/32		0.59	0.594	0.5938
				7/64	0.11	0.109	0.1094					39/64	0.61	0.609	0.6094
	1/8				0.12	0.125	0.1250		5/8				0.62	0.625	0.6250
				9/64	0.14	0.141	0.1406					41/64	0.64	0.641	0.6406
			5/32		0.16	0.156	0.1562				21/32		0.66	0.656	0.6562
				11/64	0.17	0.172	0.1719					43/64	0.67	0.672	0.6719
		3/16			0.19	0.188	0.1875			11/16			0.69	0.688	0.6875
				13/64	0.20	0.203	0.2031					45/64	0.70	0.703	0.7031
			7/32		0.22	0.219	0.2188				23/32		0.72	0.719	0.7188
				15/64	0.23	0.234	0.2344					47/64	0.73	0.734	0.7344
1/4					0.25	0.250	0.2500	3/4					0.75	0.750	0.7500
				17/64	0.27	0.266	0.2656					49/64	0.77	0.766	0.7656
			9/32		0.28	0.281	0.2812				25/32		0.78	0.781	0.7812
				19/64	0.30	0.297	0.2969					51/64	0.80	0.797	0.7969
		5/16			0.31	0.312	0.3125			13/16			0.81	0.812	0.8125
				21/64	0.33	0.328	0.3281					53/64	0.83	0.828	0.8281
			11/32		0.34	0.344	0.3438				27/32		0.84	0.844	0.8438
				23/64	0.36	0.359	0.3594					55/64	0.86	0.859	0.8594
	3/8				0.38	0.375	0.3750		7/8				0.88	0.875	0.8750
				25/64	0.39	0.391	0.3906					57/64	0.89	0.891	0.8906
			13/32		0.41	0.406	0.4062				29/32		0.91	0.906	0.9062
				27/64	0.42	0.422	0.4219					59/64	0.92	0.922	0.9219
		7/16			0.44	0.438	0.4375			15/16			0.94	0.938	0.9375
				29/64	0.45	0.453	0.4531					61/64	0.95	0.953	0.9531
			15/32		0.47	0.469	0.4688				31/32		0.97	0.969	0.9688
				31/64	0.48	0.484	0.4844					63/64	0.98	0.984	0.9844
1/2					0.50	0.500	0.5000	1					1.00	1.000	1.0000

*Omit zero to left of decimal point where used on drawings